Preface

Welcome to the fascinating world of elementary number theory, a branch of mathematics that deals with the properties and relationships of numbers, especially the integers. This book is designed to introduce readers to the beauty and utility of number theory through a collection of problems and their solutions. Whether you are a student, educator, or enthusiast, this book aims to enhance your understanding of number theory in an engaging and accessible manner.

Number theory is often considered the queen of mathematics due to its ancient origins and pure nature. It is a field that has captivated mathematicians for centuries, from the early discoveries of Euclid and Diophantus to the groundbreaking work of Fermat, Euler, Gauss, and beyond. The allure of number theory lies not only in its rich history but also in its applicability to modern-day cryptography, computer science, and problem-solving.

This book is structured to cater to a wide audience, from beginners to those with some background in number theory. The problems presented range from simple exercises that reinforce basic concepts to more challenging questions that stimulate deeper thinking and exploration. Each problem is carefully selected and crafted to illustrate a particular aspect of number theory, with solutions provided to guide the reader through the thought process and techniques involved.

The journey through this book begins with the fundamentals of divisibility, primes, and the Euclidean algorithm, gradually progressing to more complex topics such as congruences, and Diophantine equations. Along the way, readers will encounter famous theorems and conjectures that have shaped the field, including the Fundamental Theorem of Arithmetic, Fermat's Little Theorem, and Wilson's theorem.

Our goal is not only to solve problems but also to cultivate a deeper appreciation for the elegance of mathematical concepts. We encourage readers to actively engage with the material, to explore beyond the solutions provided, and to discover the joy of uncovering truths for themselves.

In preparing this book, we have drawn upon a wealth of sources, including classic texts and contemporary research, to ensure a comprehensive and enriching experience. We are grateful to the many mathematicians whose work has inspired and informed these pages, and we hope to pass on a fraction of their passion for number theory to our readers.

Whether you are embarking on this journey out of curiosity, academic interest, or the pursuit of mathematical beauty, we welcome you. May this book serve as a valuable resource and companion as you explore the intriguing world of elementary number theory.

Table of Contents

Chapter 1

Problems

1.1 Integers

Problem 1. Prove that the sum of any number of even numbers is even.

Problem 2. Prove that $\sqrt{2}$ is an irrational number.

Problem 3. Prove that $\sqrt{p}$ is irrational for any prime p.

Problem 4. For any positive integer m such that $\sqrt[n]{m}$ is a rational number, prove that $\sqrt[n]{m}$ is an integer.

Problem 5. 1. Prove that $n < 2^n$ for all integers $n \geq 2$.

2. Prove that $\sqrt[n]{n}$ is irrational for all integers $n \geq 2$.

1.2 Divisibility

Problem 6. Given a and b are two integers. Prove that $a - b \mid a^n - b^n$ for all nonnegative integers n.

Problem 7. For all positive integers n, prove the following statements:

1. $7^n - 3^n$ is divisible by 4;

2. $11^n - 5^n$ is divisible by 6;

1

3. $3^{2n} - 2^{2n}$ is divisible by 5.

Problem 8. Given a and b are two integers. Prove that $a + b \mid a^n + b^n$ for all positive odd integers n.

Problem 9. For all positive odd integers n, prove the following statements:

1. $1 + 7^n$ is divisible by 8;

2. $2^n + 2022^n$ is divisible by 2024;

3. $3^{2n} + 4^{2n}$ is divisible by 25.

Problem 10. For all positive integers n, prove that

$$2013^n + 2014^n - 2023^n - 2024^n$$

is divisible by 10.

Problem 11. For all positive odd integers n, prove that

$$1^n + 2^n + 3^n + 4^n + 5^n + 6^n$$

is divisible by 7.

Problem 12. Given m is an odd number. Prove that $1^m + 2^m + \ldots + n^m$ is divisible by $1 + 2 + \ldots + n$.

Problem 13. Given that a, b and x are three integers such that $x \mid a$ and $x \mid b$. Prove that $x \mid ma + nb$ for all $m, n \in \mathbb{Z}$.

Problem 14. Given three integers a, b and c such that $a \mid b$ and $b \mid c$. Prove that $a \mid c$.

Problem 15. Let a, b and c be three integers such that a and c are nonzero. Prove that $a \mid b$ if and only if $ac \mid bc$.

Problem 16. Given two integers a and b such that $3 \mid a$ and $5 \mid b$. Prove that $25a^2 - 90ab + 9b^2$ is divisible by 225.

Problem 17. Given two positive integers a and b such that $3 \mid a + b$. Prove that there exists a positive integer n such that $b \mid a^4 - 81n^2$.

Problem 18. Given three positive integers a, m and n such that $a \geq 2$. Prove that $a^n - 1 \mid a^m - 1$ if and only if $n \mid m$.

Problem 19. Given m and n are two positive integers such that $m \leq n$. Prove that

$$\frac{n}{\gcd(n, m)} \mid C(n, m).$$

Problem 20. Given x and y are two odd integers. Prove that $x^2 + 2y^2$ is not a perfect square.

Problem 21. For all integers a, prove that $a^2 + 1$ is never divisible by 3.

Problem 22. For all integers a, prove that $a^4 + 1$ is never divisible by 5.

Problem 23. Find all integers a such that $a^7 + 1$ is divisible by 7.

Problem 24. Find all integers a such that $a^{10} + 1$ is divisible by 10.

Problem 25. Given m and n are two positive integers such that $m > n$. Prove that $a^{2^n} + 1$ divides $a^{2^m} - 1$ for all integers a.

1.3 Modulo Congruence

Problem 26. For all positive integers n, prove the following statements:

1. $5^n - 2^n$ is divisible by 3;

2. $9^n + 16^n + 25^n - 2$ is divisible by 4;

3. $1^n + 5^n + 9^n + 13^n$ is divisible by 4;

4. $1000^n + 1001^n - (-1)^n - 1$ is divisible by 3.

Problem 27. For all positive integers n, prove the following statements:

1. $9^n - 2^n$ is divisible by 7;

2. $4^{2n} + 5^{2n} + 1$ is divisible by 3;

3. $2^n - 5^n + 7^n - 10^n$ is divisible by 3;

4. $6^{2n+1} + 7^{2n+1} + 8^{2n+1} - 1$ is divisible by 5.

3

Problem 28. Given a and b are two integers and n is a positive integers. Prove that $(a+b)^n \equiv nab^{n-1} + b^n \pmod{a^2}$.

Problem 29. For all positive integers $n \geq 2$, prove the following statements:

1. $4^n - 3n - 1 \equiv 0 \pmod 9$;

2. $5^n - 4n - 1 \equiv 0 \pmod{16}$;

3. $6^n - 5n - 1 \equiv 0 \pmod{25}$;

4. $16 \times 4^n + 9 \times 5^n - 84n - 25 \equiv 0 \pmod{144}$.

Problem 30. Given a and b are two integers. For all positive integers $n \geq 2$, prove that

$$(a+b)^n \equiv \frac{n(n-1)}{2}a^2b^{n-2} + nab^{n-1} + b^n \pmod{a^3}.$$

Problem 31. For all positive integers $n \geq 3$, prove the following statements:

1. $2 \times 4^n - 9n^2 + 3n - 2$ is divisible by 27;

2. $5^n - 8n^2 + 4n - 1$ is divisible by 64;

3. $2 \times 6^n - 25n^2 + 15n - 2$ is divisible by 125.

Problem 32. Prove that $7^{19} + 8^{19} + 9^{19} + 10^{19} + 11^{19} + 12^{19}$ is divisible by 19.

Problem 33. Given x and y are two positive integers and p is an odd prime. Prove that

$$(x+y)^{p-1} \equiv \sum_{k=1}^{p-1} x^{p-1-k}y^k \pmod p.$$

Problem 34. Given $3n$ integers $x_1, x_2, ..., x_{3n}$ which are not divisible by 3. Prove that $x_1^2 + x_2^2 + ... + x_{3n}^2$ is divisible by 3.

Problem 35. Given four integers a, b, c, d and two positive integers m and $n \geq 2$ such that $a \equiv b \pmod m$ and $c \equiv d \pmod n$. Prove that $ac + bd \equiv ad + bc \pmod{mn}$.

Problem 36. Given two positive integers a and b such that $a > b$. Prove that

$$(a-1)(a-2)\ldots(a-b) \equiv (-1)^b b! \pmod{a}.$$

Problem 37. Let p be a prime number. Prove that $p \mid C(p,k)$ for all positive integer k such that $1 \le k < p$.

Problem 38. Let p be a prime number. Prove that $p \mid C(2p,p) - p$.

Problem 39. Prove that perfect squares are of the form $4k$ or $4k + 1$, where k is a nonnegative integer.

Problem 40. For all positive integers n, prove that

$$n^4 + 2n^3 + 2n^2 + 2n + 1$$

is never a perfect square.

Problem 41. Given that p_n is the nth prime number. That is, $p_1 = 2, p_2 = 3, p_3 = 5, \ldots$ Prove that $H_n = p_1 p_2 p_3 \ldots p_n + 1$ is never a perfect square for all positive integers n.

Problem 42. Find all positive integers n such that

$$n^4 + 4n^3 + 7n^2 + 12n + 12$$

is a perfect square.

1.4 Fibonacci's Numbers

A Fibonacci sequence $\{F_n\}$ is a sequence such that $F_1 = F_2 = 1$ and $F_{n+2} = F_{n+1} + F_n$ for all $n \ge 1$. F_n is said to be the nth Fibonacci number.

Problem 43. (Cassini's Formula)
Let $\{F_n\}$ be a Fibonacci sequence. For all positive integers $n \ge 2$, prove that
$$F_{n+1} F_{n-1} - F_n^2 = (-1)^n.$$

Problem 44. Prove that any two consecutive Fibonacci numbers are relatively prime.

5

1.5 Fermat's Numbers

Fermat's numbers are numbers of the form $f_n = 2^{2^n} + 1$, where $n \geq 0$.

Problem 45. For all $n \geq 1$, prove that

$$f_0 f_1 \cdots f_{n-1} = f_n - 2.$$

Problem 46. Given two distinct nonnegative integers m and n. Prove that f_m and f_n are relatively prime.

1.6 Fermat's Theorem

Problem 47. For all prime numbers p, prove that p divides

$$5^p + 3^p + 2^p - 10.$$

Problem 48. Given three integers a, b, c and a prime number p such that $\gcd(a, p) = \gcd(b, p) = \gcd(c, p) = 1$ and $p \mid a + b + c$. Prove that p divides $ab^{p-1} + bc^{p-1} + ca^{p-1}$.

Problem 49. For all integers a, prove that 30 divides $a^{10} - a^2$.

Problem 50. For all integers a, prove that 105 divides $a^{13} - a$.

Problem 51. For all integers a, prove that 119 divides $a^{100} - a^4$.

Problem 52. Given that a is an integer such that $\gcd(a, 15) = 1$. Prove that 15 divides $a^{12} - 1$.

Problem 53. Given that a is an integer such that $\gcd(a, 14) = 1$. Prove that 28 divides $a^{12} - 1$.

Problem 54. Given a is an integer and p, q are two distinct prime numbers such that $\gcd(a, pq) = 1$. Prove that

$$p^2 a^{2(q-1)} + q^2 a^{2(p-1)} \equiv p^2 + q^2 \pmod{pq}.$$

Problem 55. Given a is an integer such that $\gcd(a, 30) = 1$. Prove that $a^4 - 1$ is divisible by 60.

Problem 56. Given p and q are two prime numbers such that $p > q$. Prove that q divides $3^p - 2^q - 3^{p-q+1} + 2$.

6

Problem 57. Suppose that p and q are two distinct prime numbers such that $a^p \equiv a \pmod{q}$ and $a^q \equiv a \pmod{p}$. Prove that

$$a^{pq} \equiv a \pmod{pq}.$$

Problem 58. Given two distinct prime numbers p and q. Prove that
$$p^{q-1} + q^{p-1} \equiv 1 \pmod{pq}.$$

Problem 59. Let a and b be positive integers. Let p be a prime number such that $\gcd(a,p) = \gcd(b,p) = 1$. Prove the following statements:

1. If $a^p \equiv b^p \pmod{p}$, then $a \equiv b \pmod{p}$.

2. If $a^p \equiv b^p \pmod{p}$, then $a^p \equiv b^p \pmod{p^2}$.

Problem 60. Let a and b be two positive integers and p be a prime number. Prove that $(a+b)^p \equiv a^p + b^p \pmod{p}$.

Problem 61. Let p be a prime number and x and y be two integers. Suppose that $x \equiv y \pmod{p-1}$. Prove that $a^x \equiv a^y \pmod{p}$ for all positive integers a such that $\gcd(a,p) = 1$.

Problem 62. Given p is a prime number and a is an integer such that $\gcd(a,p) = 1$. Prove that a^{p-2} is the inverse of a modulo p.

Problem 63. Let $p > 3$ be a prime number. Prove that

$$a^p \equiv a \pmod{6p}.$$

Problem 64. Suppose that p and q are two distinct odd prime numbers such that $p - 1 \mid q - 1$. If $\gcd(a, pq) = 1$, prove that

$$a^{q-1} \equiv 1 \pmod{pq}.$$

Problem 65. Let n be a positive integer. Suppose that $3^n - 2^n$ is a power of prime number. Prove that n is a prime number.

Problem 66. Evaluate the following expressions:

1. $\varphi(4)$

2. $\varphi(6)$

3. $\varphi(10)$

4. $\varphi(12)$

5. $\varphi\,(16)$

7. $\varphi\,(20)$

6. $\varphi\,(18)$

8. $\varphi\,(35)$

Problem 67. Find the last digits of the following numbers:

1. 3^{2025};

2. 7^{2026};

3. 13^{2027};

4. 17^{2028}.

Problem 68. Find the remainder when

1. 4^{7777} is divided by 17;

2. 5^{8888} is divided by 17;

3. 6^{9999} is divided by 17;

4. 7^{6666} is divided by 17.

Problem 69. Let $n_1, n_2, ..., n_k$ be k positive integers and a be an integer such that $\gcd(n_i, a) = 1$ for all $1 \le n \le k$. Prove that

$$a^{\mathrm{lcm}(\phi(n_1),\phi(n_2),...,\phi(n_k))} \equiv 1 \quad (\mathrm{mod}\ \mathrm{lcm}(n_1, n_2, ..., n_k)).$$

Problem 70. For all positive integers $n > 2$, prove that $\varphi(n)$ is an even number.

Problem 71. Given n is a positive integer. Prove that the sum of the positive integers less than n and relatively prime to n is $\frac{1}{2}n\varphi\,(n)$.

Problem 72. Let p be an odd prime number and a be an integer such that $\gcd(a, p) = 1$. Prove that

$$a^{\frac{p-1}{2}} \equiv \pm 1 \quad (\mathrm{mod}\ p).$$

Problem 73. Find all positive integers x, y and z such that

$$\varphi\,(x - 1) + \varphi\,(y - 2) + \varphi\,(z - 3) = 3.$$

Problem 74. Given a and b are two positive integers such that $\gcd(a, b) = 1$. Prove that $a^{\phi(b)} + b^{\phi(a)} \equiv 1\ (\mathrm{mod}\ ab)$.

Problem 75. Given $a \ge 2$ is an integer and n is a positive integer. Prove that $\varphi\,(a^n - 1)$ is divisible by n.

Problem 76. Given p and q are two distinct prime numbers. Let a be an integer such that $\gcd(a, p) = \gcd(a, q) = 1$. Prove that

$$p^2 a^{q(q-1)} + q^2 a^{p(p-1)} \equiv p^2 + q^2 \quad (\mathrm{mod}\ p^2 q^2).$$

1.7 Wilson's Theorem

Problem 77. Find the remainder when 85! is divided by 89.

Problem 78. Find the remainder when 104! is divided by 109.

Problem 79. Prove that $(41!)^2 - 1517 \times 40! + 38 \equiv 0 \pmod{2021}$.

Problem 80. Suppose that p is a prime number. Prove that

$$(p-2)! \equiv 1 \pmod{p}.$$

Problem 81. Let p be an odd prime. Prove the following statements:

$$1^2 \times 3^2 \times 5^2 \times ... \times (p-2)^2 \equiv (-1)^{\frac{p+1}{2}} \pmod{p}.$$

Problem 82. Is $171 \times 534! - 1$ a composite number?

Problem 83. (Lerch's Congruence)
Given p is an odd prime. Prove that

$$1^{p-1} + 2^{p-1} + ... + (p-1)^{p-1} \equiv p + (p-1)! \pmod{p^2}.$$

Problem 84. Given a prime number p and an integer n such that $0 \le n \le p-1$. Prove that

$$n!\,(p-n-1)! + (-1)^n \equiv 0 \pmod{p}.$$

Problem 85. Given x is an integer such that

$$1 + \frac{1}{2} + \frac{1}{3} + ... + \frac{1}{2017} = \frac{x}{2017!}.$$

Prove that $x \equiv 1 \pmod{1009}$.

Problem 86. Given a prime number p. Prove that

$$(p-1)! \equiv p-1 \pmod{p(p-1)}.$$

Problem 87. Given two positive integers a and b such that $a \ge 2$ and $\gcd(a,b) = 1$. Prove that

$$b^{a-1} + (a-1)! \equiv 0 \pmod{a}$$

if and only if a is a prime number.

9

1.8   The Greatest Common Divisor and Least Common Multiple of Positive Integers

Problem 88. Find the greatest common divisor of the following integers:

1. 18 and 36   3. 45 and 57   5. 200 and 150

2. 20 and 30   4. 108 and 99   6. 198 and 196

Problem 89. Prove that there are no integers x and y such that $x + y = 11$ and $\gcd(x, y) = 4$.

Problem 90. Let a, b and k be two positive integers. Suppose that $\gcd(\gcd(a, a + kb), k) = 1$. Prove that

$$\gcd(a, a + kb) = \gcd(a, b).$$

Problem 91. A linear combination of two positive integers a and b is the sum of the multiples of a and b. That is, the linear combination of two positive integers a and b is an integer of the form $ax + by$, where x and y are integers. Prove that $\gcd(a, b)$ is a linear combination of a and b.

Problem 92. Suppose that a and b are two positive integers. Let $d = \gcd(a, b)$ and d' is a common divisor of a and b. Prove that $d' \mid d$.

Problem 93. Let a and b be two positive integers. Suppose that $a = dm$ and $b = dn$. Prove that $d = \gcd(a, b)$ if and only if $\gcd(m, n) = 1$.

Problem 94. Given a, b and k are three positive integers. Prove that

$$\gcd(ak, bk) = k \gcd(a, b).$$

Problem 95. Given two positive integers a and b. Prove that a and b are relatively prime if and only if $ax + by = 1$ for some integers x and y.

Problem 96. Prove that every two consecutive positive integers are relatively prime.

Problem 97. Prove that the following fractions are irreducible for all positive integers n:

1.8. The Greatest Common Divisor and Least Common Multiple of Positive Integers

1. $\dfrac{n+1}{7n+6}$

2. $\dfrac{n(n+2)}{n+1}$

3. $\dfrac{n+1}{n^2+4n+2}$

4. $\dfrac{n+1}{n(n^2+3n+3)}$

Problem 98. Given a, b and c are three positive integers such that $\gcd(a,b) = 1$ and $\gcd(a,c) = 1$. Prove that $\gcd(a, bc) = 1$.

Problem 99. Suppose that a, b and c are three positive integers such that $\gcd(b,c) = 1$. Prove that $\gcd(a,b) = \gcd(ac,b)$.

Problem 100. Given two positive integers a and b such that $\gcd(a,b) = 1$. Prove that $\gcd(a^n, b^n) = 1$ for all positive integers n.

Problem 101. Given two positive integers a and b such that

$$\gcd(a,b) = 1.$$

Prove that $\gcd(a + b, ab) = 1$.

Problem 102. Given two positive integers a and b such that

$$\gcd(a,b) = 1.$$

Prove that

1. $\gcd\left(a+b, b^2\right) = 1$; 2. $\gcd\left(a+b, 3a^2+b^2\right) = 1$.

Problem 103. Given a, b, k and l are four positive integers such that $\gcd(a,b) = 1$ and $k + l = 2^m$ for some nonnegative integers m. Prove that $\gcd(a + b, ka^2 + lb^2) = 1$.

Problem 104. Given two positive integers a and b such that

$$\gcd(a,b) = 1.$$

Prove that

1. $\gcd\left(a^2 + 3b^2, (a+b)^3 + 8b^3 + a + b\right) = 1$;

2. $\gcd\left(a^2 + 3b^2, a^4 + 4a^2b^2 + 3b^4 + a + b\right) = 1$.

Problem 105. Given two positive integers a and b such that

$$\gcd(a,b) = 1.$$

Prove that

11

1. $\gcd(a+b, a-b) = 1$ or 2;   2. $\gcd(a+b, 3a+2b) = 1$.

Problem 106. Given four positive integers a, b, x and y such that $\gcd(a,b) = \gcd(x,y)$. Prove that

$$\gcd\left(a^2, b^2\right) = \gcd\left(x^2, y^2\right).$$

Problem 107. Given a and b are two positive integers. Let x and y be two integers such that $ax + by = m$. If $d = \gcd(a,b)$, prove that $d \mid m$.

Problem 108. Given n is a positive integer. Prove that

$$\gcd\left(n! + 1, (n+1)! + 1\right) = 1.$$

Problem 109. Given three positive integers a, x, y such that $\gcd(x,y) = 1$ and $a \mid xy$. Prove that $a = bc$ for some positive integers b and c such that $\gcd(b,c) = 1$.

Problem 110. Given three positive integers a, b and c. Prove that

1. $\gcd(a, b, c) = \gcd\left[\gcd(a, b), c\right]$;

2. $\operatorname{lcm}(a, b, c) = \operatorname{lcm}\left[\operatorname{lcm}(a, b), c\right]$.

Problem 111. Given x and y are two positive integers such that $\gcd(x,y) = 1$. Prove that $\operatorname{lcm}(x,y) = xy$.

Problem 112. Given three positive integers a, b and c such that $a \mid c$ and $b \mid c$. Prove that $\operatorname{lcm}(a,b) \mid c$.

Problem 113. Given three positive integers a, b and x such that $a \mid x, b \mid x$ and $\gcd(a,b) = 1$. Prove that $ab \mid x$.

Problem 114. Given five positive integers a, b, c, x and y such that $\gcd(a,b) = 1$. Suppose that $a \mid c^x - 1$ and $b \mid c^y - 1$. Prove that

$$ab \mid c^{\operatorname{lcm}(x,y)} - 1.$$

Problem 115. Given three positive integers a, x and y such that $\gcd(x,y) = 1$. Prove that $\gcd(a, xy) = \gcd(a, x) \times \gcd(a, y)$.

Problem 116. Given a and b are two positive integers. If $\gcd(a,b) = 3$, find the following expressions:

1. $\gcd\left(a^3, b^2\right)$;
3. $\gcd\left(\gcd\left(a^2, b^2\right), 2a^2\right)$;

2. $\gcd\left(2a^2, b^3\right)$;
4. $\gcd\left(\gcd\left(a^2, b^2\right), 3b^3\right)$.

Problem 117. Given two positive integers a and b. Let d be a positive odd number such that $d \mid a + b$ and $d \mid a - b$. Prove that

$$d \mid \gcd(a, b).$$

Problem 118. Given positive integers a, b, m, n, x and y, where b is a prime number. Suppose that $d \mid ax + by, d \mid am + bn$ and $my - nx = b$. Find the possible values of $\gcd\left(ax + by, am + bn\right)$.

Problem 119. Given n and d are two positive integers such that $d \mid 3n + 1$ and $d \mid 5n + 2$. Find d.

Problem 120. Given n and d are two positive integers such that $d \mid 5n - 2$ and $d \mid 4n - 3$. Find d.

Problem 121. Given a and b are positive integers such that $\gcd(a, b) = 1$ and ab is a perfect square. Prove that a and b are perfect squares.

Problem 122. Given x, y and z are three integers such that $\gcd(x, y) = \gcd(y, z) = \gcd(z, x) = 1$. Prove that

$$\gcd\left(x, y, z\right) \times \operatorname{lcm}\left(x, y, z\right) = xyz.$$

Problem 123. Given two positive integers a and b such that $\gcd(a, b) = 1$. Find the possible values of the following expressions:

1. $\gcd\left(a, a + 1, a + 2\right)$;

2. $\gcd\left(a, (a + 1)(a + 2)\right)$;

3. $\gcd\left(a, (a + 2)(a, +3)\right)$;

4. $\gcd\left(a(a + 1), (a + 2)(a + 3)\right)$;

5. $\gcd\left(a^2 - b^2, a^2 + b^2\right)$;

6. $\gcd\left(2a - 2b, a^3 - b^3\right)$;

7. $\gcd\left(a^2 - b^2, a^3 - b^3\right)$.

Problem 124. For all positive integers a, prove that

$$\gcd\left(a^{2^n} + 1, a^{2^m} + 1\right) = 1 \text{ or } 2$$

for all positive integers m and n.

Problem 125. Given a positive integer n. Prove that

$$\frac{\gcd\left(n\left(n+1\right)\left(n+2\right), 3n+9\right)}{\gcd\left(n\left(n+2\right), \left(n+1\right)\left(n+3\right)\right)}$$

is an integer.

Problem 126. Given a positive integer n. Simplify

$$F = \frac{\gcd\left(n\left(n+1\right), \left(n+1\right)\left(n+3\right)\right)}{\gcd\left(n\left(n+1\right), \left(n+1\right)^2\right)}.$$

Problem 127. Let n be an even number. Find all positive integers a and b such that $\gcd(a, b) = 1$ and $a + b \mid a^n + b^n$.

Problem 128. Given a and b are two distinct positive integers. Prove that

$$|a - b| \geq \gcd(a, b).$$

Problem 129. Given that a and b are two positive integers. Prove that

$$\operatorname{lcm}\left(a, b\right) \times \gcd\left(a, b\right) = ab.$$

Problem 130. Given a and b are two positive integers such that $a > b$. Prove that

$$\frac{1}{\operatorname{lcm}\left(a, b\right)} \leq \frac{1}{b} - \frac{1}{a}.$$

Problem 131. Given $x_1, x_2, ..., x_n$ is a strictly increasing sequence of positive integers. Prove that

$$\sum_{k=1}^{n-1} \frac{1}{\operatorname{lcm}\left(x_k, x_{k+1}\right)} \leq 1 - \frac{1}{x_n}.$$

Problem 132. Given two positive integers a and b such that $a > b$ and $\dfrac{\operatorname{lcm}\left(a, b\right)}{\gcd\left(a, b\right)} = a - b$. Prove that ab is a perfect cube.

1.8. The Greatest Common Divisor and Least Common Multiple of Positive Integers

Problem 133. For all positive integers n, prove that

$$\frac{n!}{(x+1)(x+2)\dots(x+n)} = \sum_{k=1}^{n} \frac{(-1)^{k-1}kC(n,k)}{x+k}.$$

Problem 134. Given two positive integers x and y. Prove that $yC(x+y,y)$ divides $\operatorname{lcm}(x+1, x+2, ..., x+y)$.

Problem 135. Given $x_1, x_2, ..., x_n$ are n positive integers. Prove that
$$\operatorname{lcm}(x_1, x_2, ..., x_n) \geq \frac{x_1 + x_2 + ... + x_n}{n}.$$

Problem 136. For all positive integers n greater than 1, prove that

$$\frac{\operatorname{lcm}(1, 2, ..., n+1)}{n+1} = \operatorname{lcm}(C(n,0), C(n,1), ..., C(n,n)).$$

Problem 137. For all positive integers $n \geq 2$, prove that

$$\operatorname{lcm}(1, 2, 3, ..., n) \geq 2^{n-1}.$$

Problem 138. Given three positive integers x, y and z. Prove that $\gcd(x, yz)$ divides $\gcd(x, y) \times \gcd(x, z)$.

Problem 139. Given $x_1, x_2, ..., x_n$ are n positive integers. Prove that

$$\gcd(x_1 x_2 ... x_{n-1}, x_n) \leq \prod_{k=1}^{n-1} \gcd(x_k, x_n)$$

for all $n \geq 3$.

Problem 140. Given n positive integers $x_1, x_2, ..., x_n$. Prove that

$$\operatorname{lcm}(x_1, x_2, ..., x_n) \geq \frac{x_1 x_2 ... x_n}{\prod_{1 \leq i < j \leq n} \gcd(x_i, x_j)}.$$

Chapter 2

Solutions

Problem 1. Prove that the sum of any number of even numbers is even.

Solution. Let $n_1, n_2, ..., n_k$ be even numbers. We shall show that $n_1 + n_2 + ... + n_k$ is an even number. Since $n_1, n_2, ..., n_k$ are even numbers, then $n_1 = 2m_1, n_2 = 2m_2, ..., n_k = 2m_k$. It follows that

$$n_1 + n_2 + ... + n_k = 2m_1 + 2m_2 + ... + 2m_k$$
$$= 2(m_1 + m_2 + ... + m_k)$$
$$= 2m$$

, where $m = m_1 + m_2 + ... + m_k$ is an integer.
Therefore, the statement is proved.

Problem 2. Prove that $\sqrt{2}$ is an irrational number.

Solution. We prove the given statement by contradiction. Suppose that $\sqrt{2}$ is a rational number. That is, $\sqrt{2} = \dfrac{a}{b}$, where a and b are relatively prime.

Squaring both sides, we obtain $2 = \dfrac{a^2}{b^2}$ or $a^2 = 2b^2$.
It follows that $2 \mid a^2$. Then $2 \mid a$.
It implies that $a = 2k$ for some integer k.
Then $(2k)^2 = 2b^2$ or $b^2 = 2k^2$.
Hence, $2 \mid b^2$. Then $2 \mid b$.
We obtain 2 is a common divisor of a and b, which is a contradiction.
Therefore, $\sqrt{2}$ is an irrational number.

Problem 3. Prove that $\sqrt{p}$ is irrational for any prime p.

Solution. We prove the above statement by contradiction. Suppose that $\sqrt{p}$ is a rational number. Then $\sqrt{p} = \dfrac{a}{b}$, where a and b are positive integers such that $\gcd(a, b) = 1$.

We obtain $p = \dfrac{a^2}{b^2}$ or $a^2 = pb^2$.

It follows that $p \mid a^2$. Then $p \mid a$.

Hence, there exists an integer k such that $a = kp$.

We obtain $(kp)^2 = pb^2$ or $b^2 = k^2p$.

Then $p \mid b^2$. Hence, $p \mid b$.

Thus, p is a common divisor of a and b. This leads to a contradiction which saying that a and b are relatively prime.

Therefore, $\sqrt{p}$ is an irrational number for any prime number p.

Problem 4. For any positive integer m such that $\sqrt[n]{m}$ is a rational number, prove that $\sqrt[n]{m}$ is an integer.

Solution. Since $\sqrt[n]{m} = \dfrac{a}{b}$, where a and b are integers such that

$$\gcd(a, b) = 1.$$

Let $a = p_1 p_2 ... p_r$ and $b = q_1 q_2 ... q_s$, where $p_i \neq q_j$ for all i and j.

From $\sqrt[n]{m} = \dfrac{a}{b}$, we obtain $m = \dfrac{a^n}{b^n}$ or $a^n = mb^n$.

It follows that

$$(p_1 p_2 ... p_r)^n = m(q_1 q_2 ... q_s)^n.$$

Then $(p_1 p_2 ... p_r)^n \mid m$.

It turns out that there exists an integer k such that $m = k(p_1 p_2 ... p_r)^n$.

We obtain

$$(p_1 p_2 ... p_r)^n = k(p_1 p_2 ... p_r)^n (q_1 q_2 ... q_s)^n$$

or

$$k(q_1 q_2 ... q_s)^n = 1.$$

Hence, $q_1 = q_2 = ... = q_s = 1$. We obtain $b = 1$.

Therefore, $\sqrt[n]{m}$ is an integer.

Problem 5.   1. Prove that $n < 2^n$ for all integers $n \geq 2$.

2. Prove that $\sqrt[n]{n}$ is irrational for all integers $n \geq 2$.

Solution. 1. Prove that $n < 2^n$ for all integers $n \geq 2$.
We will prove the statement by using induction.
For $n = 1$, we obtain $n < 2^n$ is true.
Suppose that $n < 2^n$. We shall show that $n + 1 < 2^{n+1}$.
From $n < 2^n$, it follows that

$$n + 1 < 2^n + 1 < 2^n + 2^n = 2 \times 2^n = 2^{n+1}.$$

Therefore, $n < 2^n$ for all integers $n \geq 2$.

2. Prove that $\sqrt[n]{n}$ is irrational for all integers $n \geq 2$.
Suppose that $\sqrt[n]{n}$ is rational. From Problem 4, we obtain $\sqrt[n]{n}$ is an integer. Then $\sqrt[n]{n} = m$ for some integer m.
It follows that $n = m^n$.
From the above proof, we have $n < 2^n$ for all integer $n \geq 2$.
It implies that $m^n < 2^n$. We obtain $m < 2$. Hence, $m = 1$.
If $m = 1$, we obtain $n = 1^n = 1$, which is a contradiction to the assumption that $n \geq 2$.
Therefore, $\sqrt[n]{n}$ is irrational for all integers $n \geq 2$.

Problem 6. Given a and b are two integers. Prove that $a - b \mid a^n - b^n$ for all nonnegative integers n.

Solution. We have

$$a^n - b^n = (a - b)\left(a^{n-1} + a^{n-2}b + \ldots + b^{n-1}\right)$$
$$= (a - b)\,k$$

, where $k = a^{n-1} + a^{n-2}b + \ldots + b^{n-1}$ is an integer.
Therefore, $a - b \mid a^n - b^n$ for all nonnegative integers n.

Problem 7. For all positive integers n, prove the following statements:

1. $7^n - 3^n$ is divisible by 4;

2. $11^n - 5^n$ is divisible by 6;

3. $3^{2n} - 2^{2n}$ is divisible by 5.

Solution. For all positive integers n, prove the following statements:

1. $7^n - 3^n$ is divisible by 4;
   From the previous problem, we have $7 - 3 \mid 7^n - 3^n$.
   It follows that $4 \mid 7^n - 3^n$.
   Therefore, $7^n - 3^n$ is divisible by 4.

2. $11^n - 5^n$ is divisible by 6;
   From the previous problem, we have $11 - 5 \mid 11^n - 5^n$.
   It follows that $6 \mid 11^n - 5^n$.
   Therefore, $11^n - 5^n$ is divisible by 6.

3. $3^{2n} - 2^{2n}$ is divisible by 5.
   We have $3^{2n} - 2^{2n} = 9^n - 4^n$.
   It follows that $9 - 4 \mid 3^{2n} - 2^{2n}$.
   Then $5 \mid 3^{2n} - 2^{2n}$.
   Therefore, $3^{2n} - 2^{2n}$ is divisible by 5.

Problem 8. Given a and b are two integers. Prove that $a + b \mid a^n + b^n$ for all positive odd integers n.

Solution. For all positive odd integers n, we have

$$a^n + b^n = (a + b)\left(a^{n-1} - a^{n-2}b + ... + b^{n-1}\right)$$
$$= (a + b)\,k$$

, where $k = a^{n-1} - a^{n-2}b + ... + b^{n-1}$ is an integer.
Therefore, $a + b \mid a^n + b^n$ for all positive odd integers n.

Problem 9. For all positive odd integers n, prove the following statements:

1. $1 + 7^n$ is divisible by 8;

2. $2^n + 2022^n$ is divisible by 2024;

3. $3^{2n} + 4^{2n}$ is divisible by 25.

Solution. For all positive odd integers n, prove the following statements:

1. $1 + 7^n$ is divisible by 8;
   We have $1 + 7^n = 1^n + 7^n$.
   From the previous problem, we have $1 + 7 \mid 1^n + 7^n$.
   Then $8 \mid 1 + 7^n$.
   Therefore, $1 + 7^n$ is divisible by 8.

2. $2^n + 2022^n$ is divisible by 2024;

From the previous problem, we have $2 + 2022 \mid 2^n + 2022^n$.

Then $2024 \mid 2^n + 2022^n$.

Therefore, $2^n + 2022^n$ is divisible by 2024.

3. $3^{2n} + 4^{2n}$ is divisible by 25.

We have $3^{2n} + 4^{2n} = 9^n + 16^n$.

From the previous problem, we have $9 + 16 \mid 9^n + 16^n$.

Then $25 \mid 9^n + 16^n$.

Therefore, $3^{2n} + 4^{2n}$ is divisible by 25.

Problem 10. For all positive integers n, prove that

$$2013^n + 2014^n - 2023^n - 2024^n$$

is divisible by 10.

Solution. We have

$$
\begin{aligned}
&2013^n + 2014^n - 2023^n - 2024^n \\
&= (2013^n - 2023^n) + (2014^n - 2024^n) \\
&= (2013 - 2023) k_1 + (2014 - 2024) k_2 \\
&= -10k_1 - 10k_2 \\
&= 10 \left(-k_1 - k_2\right) \\
&= 10k
\end{aligned}
$$

, where $k = -k_1 - k_2$ is an integer.

Therefore, $2013^n + 2014^n - 2023^n - 2024^n$ is divisible by 10.

Problem 11. For all positive odd integers n, prove that

$$1^n + 2^n + 3^n + 4^n + 5^n + 6^n$$

is divisible by 7.

Solution. We have

$$
\begin{aligned}
&1^n + 2^n + 3^n + 4^n + 5^n + 6^n \\
&= (1^n + 6^n) + (2^n + 5^n) + (3^n + 4^n) \\
&= (1 + 6) k_1 + (2 + 5) k_2 + (3 + 4) k_3 \\
&= 7k_1 + 7k_2 + 7k_3 \\
&= 7 \left(k_1 + k_2 + k_3\right)
\end{aligned}
$$

$$= 7k$$

, where $k = k_1 + k_2 + k_3$ is an integer.
Therefore, $1^n + 2^n + 3^n + 4^n + 5^n + 6^n$ is divisible by 7.

Problem 12. Given m is an odd number. Prove that $1^m + 2^m + \ldots + n^m$ is divisible by $1 + 2 + \ldots + n$.

Solution. Let $S = 1^m + 2^m + \ldots + n^m$.
Then $S = n^m + (n-1)^m + \ldots + 1$.
It follows that

$$2S = (n^m + 1) + [(n-1)^m + 2^m] + \ldots + (n^m + 1^m)$$
$$= (n+1) k_1 + (n+1) k_2 + \ldots + (n+1) k_n$$

, where $k_1, k_2, \ldots, k_n$ are integers.
Then

$$2S = (n+1) (k_1 + k_2 + \ldots + k_n).$$

It implies that

$$n + 1 \mid 2S. \tag{1}$$

Moreover,

$$2S = [1^m + (n-1)^m] + [2^m + (n-2)^m] + \ldots + [(n-1)^m + 1] + 2n^m$$
$$= nj_1 + nj_2 + \ldots + nj_{n-1} + 2n^m$$

, where $j_1, j_2, \ldots, j_{n-1}$ are integers.
We obtain

$$2S = n \left(j_1 + j_2 + \ldots + j_{n-1} + 2n^{m-1}\right).$$

It follows that

$$n \mid 2S. \tag{2}$$

From (1) and (2), since $\gcd(n+1, n) = 1$, we obtain $n(n+1) \mid 2S$.
It turns out that

$$\frac{n(n+1)}{2} \mid S$$

since $n(n+1)$ and $2S$ are both divisible by 2.
Moreover, $\dfrac{n(n+1)}{2} = 1 + 2 + \ldots + n$.
Therefore, $1^m + 2^m + \ldots + n^m$ is divisible by $1 + 2 + \ldots + n$.

Problem 13. Given that a, b and x are three integers such that $x \mid a$ and $x \mid b$. Prove that $x \mid ma + nb$ for all $m, n \in \mathbb{Z}$.

Solution. Since $x \mid a$ and $x \mid b$, it follows that $a = kx$ and $b = lx$ for some integers k and l. We obtain

$$ma + nb = m(kx) + n(lx)$$
$$= (mk + nl)x.$$

Therefore, $x \mid ma + nb$ for all $m, n \in \mathbb{Z}$.

Problem 14. Given three integers a, b and c such that $a \mid b$ and $b \mid c$. Prove that $a \mid c$.

Solution. Since $a \mid b$ and $b \mid c$, it follows that $b = am$ and $c = nb$ for some integers m and n. We obtain $c = n(am) = (mn)a$. Therefore, $a \mid c$.

Problem 15. Let a, b and c be three integers such that a and c are nonzero. Prove that $a \mid b$ if and only if $ac \mid bc$.

Solution. $\Rightarrow$ Suppose that $a \mid b$. We shall show that $ac \mid bc$. Since $a \mid b$, it follows that $b = ak$ for some integers k. Multiply both of the equality by c, we obtain $bc = (ac)k$. It turns out that $ac \mid bc$.
$\Leftarrow$ Suppose that $ac \mid bc$. We shall show that $a \mid b$. Since $ac \mid bc$, it follows that $bc = ack$ for some integers k. Divide both sides of the equality by c, we obtain $b = ak$. Hence, $a \mid b$. Therefore, the statement is proved.

Problem 16. Given two integers a and b such that $3 \mid a$ and $5 \mid b$. Prove that $25a^2 - 90ab + 9b^2$ is divisible by 225.

Solution. Since $3 \mid a$ and $5 \mid b$, it follows that $a = 3m$ and $b = 3n$ for some integers m and n. We obtain

$$25a^2 - 90ab + 9b^2$$
$$= 25(3m)^2 - 90(3m)(5n) + 9(5n)^2$$
$$= 25(9m^2) - 25(54mn) + 25(9n^2)$$
$$= 25(9m^2 - 54mn + 9n^2)$$
$$= 25k$$

, where $k = 9m^2 - 54mn + 9n^2 \in \mathbb{Z}$.
Therefore, $25a^2 - 90ab + 9b^2$ is divisible by 225.

Problem 17. Given two positive integers a and b such that $3 \mid a + b$. Prove that there exists a positive integer n such that $b \mid a^4 - 81n^2$.

Solution. Since $3 \mid a + b$, then there exists an integer k such that $a + b = 3k$. We obtain $a = 3k - b$. Squaring both sides of the equality, we obtain

$$a^2 = (3k - b)^2$$
$$= 9k^2 - 6bk + b^2$$
$$= 9k^2 + (-6k + b)b$$

Let $l = -6k + b \in \mathbb{Z}$. It follows that $a^2 = 9k^2 + bl$.
Squaring both sides of the last equality, we obtain

$$a^4 = \left(9k^2 + bl\right)^2$$
$$= 81k^4 + 18bk^2l + b^2l^2$$
$$= 81k^4 + b\left(18k^2l + bl^2\right)$$
$$= 81k^4 + bm$$

, where $m = 18k^2l + bl^2 \in \mathbb{Z}$.
Hence, $a^4 - 81k^4 = bm$.
It turns out that $b \mid a^4 - 81k^4$.
That is, there exists a positive integer $n = k^2$ such that $b \mid a^4 - 81n^2$.

Problem 18. Given three positive integers a, m and n such that $a \geq 2$. Prove that $a^n - 1 \mid a^m - 1$ if and only if $n \mid m$.

Solution. $\Rightarrow$ Suppose that $a^n - 1 \mid a^m - 1$. We shall show that $n \mid m$. Assume that $n \nmid m$. Then $m = nq + r$, where $0 < r < n$. We obtain

$$a^m - 1 = a^{nq+r} - 1$$
$$= a^r(a^n)^q - a^r + a^r - 1$$
$$= a^r(a^n - 1)\left[a^{(n-1)q} + a^{(n-2)q} + \dots + 1\right] + a^r - 1$$
$$= a^r(a^n - 1) + a^r - 1.$$

Since $0 < r < n$, then $0 < a^r - 1 < a^n$. Hence, $a^n - 1 \nmid a^m - 1$.
$\Leftarrow$ Suppose that $n \mid m$. We shall show that $a^n - 1 \mid a^m - 1$. Since $n \mid m$, we obtain $m = nq$ for some integers q. It follows that

$$a^m - 1 = a^{nq} - 1$$

$$= (a^n)^q - 1$$
$$= (a^n - 1)\left[a^{(n-1)q} + a^{(n-2)q} + \dots + 1\right]$$
$$= (a^n - 1)k$$

, where $k = a^{(n-1)q} + a^{(n-2)q} + \dots + 1 \in \mathbb{N}$.
Hence, $a^n - 1 \mid a^m - 1$.
Therefore, $a^n - 1 \mid a^m - 1$ if and only if $n \mid m$.

Problem 19. Given m and n are two positive integers such that $m \leq n$. Prove that

$$\frac{n}{\gcd(n,m)} \mid C(n,m).$$

Solution. To prove that $\dfrac{n}{\gcd(n,m)} \mid C(n,m)$, it is sufficient to prove that

$$K = \frac{C(n,m)}{\dfrac{n}{\gcd(n,m)}} = \frac{\gcd(n,m)}{n}C(n,m)$$

is an integer. Using Bézout's lemma, there exist integers x and y such that $mx + ny = \gcd(n,m)$. It turns out that

$$K = \frac{mx + ny}{n}C(n,m)$$
$$= yC(n,m) + x\frac{m}{n}C(n,m).$$

We have

$$\frac{m}{n}C(n,m) = \frac{m}{n} \times \frac{n!}{m!(n-m)!}$$
$$= \frac{(n-1)!}{(m-1)!(n-m)!}$$
$$= C(n-1, m-1).$$

Consequently, $K = yC(n,m) + xC(n-1, m-1)$ is an integer.
Therefore, $\dfrac{n}{\gcd(n,m)} \mid C(n,m)$.

Problem 20. Given x and y are two odd integers. Prove that $x^2 + 2y^2$ is not a perfect square.

Solution. Since x is an odd integer, we obtain $x = 2k+1$ for some integers k. Then

$$
\begin{aligned}
x^2 &= (2k+1)^2 \\
&= 4k^2 + 4k + 1 \\
&\equiv 1 \quad (\text{mod } 4).
\end{aligned}
$$

Similarly, we obtain $y^2 \equiv 1 \pmod 4$.
It turns out

$$
\begin{aligned}
x^2 + 2y^2 &\equiv 1 + 2 \quad (\text{mod } 4) \\
&\equiv 3 \quad (\text{mod } 4).
\end{aligned}
$$

Therefore, $x^2 + 2y^2$ is not a perfect square.

Problem 21. For all integers a, prove that $a^2 + 1$ is never divisible by 3.

Solution. For all integers a, we obtain $a = 3k, 3k+1$ or $3k+2$ for some integers k.

- For $a = 3k$, we obtain

$$
\begin{aligned}
a^2 + 1 &= (3k)^2 + 1 \\
&= 9k^2 + 1 \\
&\equiv 1 \quad (\text{mod } 3).
\end{aligned}
$$

In this case, $a^2 + 1$ is not divisible by 3.

- For $a = 3k + 1$, we obtain

$$
\begin{aligned}
a^2 + 1 &= (3k+1)^2 + 1 \\
&\equiv 1 + 1 \quad (\text{mod } 3) \\
&\equiv 2 \quad (\text{mod } 3).
\end{aligned}
$$

In this case, $a^2 + 1$ is not divisible by 3.

- For $a = 3k + 2$, we obtain

$$
\begin{aligned}
a^2 + 1 &= (3k+2)^2 + 1 \\
&\equiv 2^2 + 1 \quad (\text{mod } 3) \\
&\equiv 5 \quad (\text{mod } 3) \\
&\equiv 2 \quad (\text{mod } 3).
\end{aligned}
$$

In this case, $a^2 + 1$ is not divisible by 3.

Therefore, $a^2 + 1$ is never divisible by 3.

Problem 22. For all integers a, prove that $a^4 + 1$ is never divisible by 5.

Solution. For all integers a, we obtain $a = 5k, 5k+1, 5k+2, 5k+3$ or $5k + 4$ for some integers k.

- For $a = 5k$, we obtain

$$a^4 + 1 = (5k)^4 + 1$$
$$\equiv 1 \pmod 5.$$

- For $a = 5k + 1$, we obtain

$$a^4 + 1 = (5k + 1)^4 + 1$$
$$\equiv 1 + 1 \pmod 5$$
$$\equiv 2 \pmod 5.$$

- For $a = 5k + 2$, we obtain

$$a^4 + 1 = (5k + 2)^4 + 1$$
$$\equiv 2^4 + 1 \pmod 5$$
$$\equiv 17 \pmod 5$$
$$\equiv 2 \pmod 5.$$

- For $a = 5k + 3$, we obtain

$$a^4 + 1 = (5k + 3)^4 + 1$$
$$\equiv 3^4 + 1 \pmod 5$$
$$\equiv 82 \pmod 5$$
$$\equiv 2 \pmod 5.$$

- For $a = 5k + 4$, we obtain

$$a^4 + 1 = (5k + 4)^4 + 1$$
$$\equiv 4^4 + 1 \pmod 5$$
$$\equiv 257 \pmod 5$$
$$\equiv 2 \pmod 5.$$

Therefore, $a^4 + 1$ is never divisible by 5.

Problem 23. Find all integers a such that $a^7 + 1$ is divisible by 7.

Solution. For all integers a, we obtain $a = 7k, 7k+1, 7k+2, 7k+3, 7k+4, 7k+5$ or $7k+6$.

- For $a = 7k$, we obtain

$$a^7 + 1 = (7k)^7 + 1$$
$$\equiv 1 \pmod 7.$$

- For $a = 7k + 1$, we obtain

$$a^7 + 1 = (7k+1)^7 + 1$$
$$\equiv 1^7 + 1 \pmod 7$$
$$\equiv 2 \pmod 7.$$

- For $a = 7k + 2$, we obtain

$$a^7 + 1 = (7k+2)^7 + 1$$
$$\equiv 2^7 + 1 \pmod 7$$
$$\equiv 129 \pmod 7$$
$$\equiv 3 \pmod 7.$$

- For $a = 7k + 3$, we obtain

$$a^7 + 1 = (7k+3)^7 + 1$$
$$\equiv 3^7 + 1 \pmod 7$$
$$\equiv 4 \pmod 7.$$

- For $a = 7k + 4$, we obtain

$$a^7 + 1 = (7k+4)^7 + 1$$
$$\equiv 4^7 + 1 \pmod 7$$
$$\equiv -3^7 + 1 \pmod 7$$
$$\equiv 5 \pmod 7.$$

- For $a = 7k + 5$, we obtain

$$\begin{aligned}
a^7 + 1 &= (7k + 5)^7 + 1 \\
&\equiv 5^7 + 1 \pmod{7} \\
&\equiv -2^7 + 1 \pmod{7} \\
&\equiv -127 \pmod{7} \\
&\equiv 6 \pmod{7}.
\end{aligned}$$

- For $a = 7k + 6$, we obtain

$$\begin{aligned}
a^7 + 1 &= (7k + 6)^7 + 1 \\
&\equiv 6^7 + 1 \pmod{7} \\
&\equiv -1^7 + 1 \pmod{7} \\
&\equiv 0 \pmod{7}.
\end{aligned}$$

Therefore, $a = 7k + 6$, where k is an integer.

Problem 24. Find all integers a such that $a^{10} + 1$ is divisible by 10.

Solution. For all integers a, we obtain $a = 10k, 10k \pm 1, 10k \pm 2, 10k \pm 3, 10k \pm 4$ or $10k \pm 5$ for some integers k.

- For $a = 10k$, we obtain

$$\begin{aligned}
a^{10} + 1 &= (10k)^4 + 1 \\
&\equiv 1 \pmod{10}.
\end{aligned}$$

- For $a = 10k \pm 1$, we obtain

$$\begin{aligned}
a^{10} + 1 &= (10k \pm 1)^{10} + 1 \\
&\equiv 1^{10} + 1 \pmod{10} \\
&\equiv 2 \pmod{10}.
\end{aligned}$$

- For $a = 10k \pm 2$, we obtain

$$\begin{aligned}
a^{10} + 1 &= (10k \pm 2)^{10} + 1 \\
&\equiv 2^{10} + 1 \pmod{10} \\
&\equiv 4 \pmod{10}.
\end{aligned}$$

- For $a = 10k \pm 3$, we obtain

$$
\begin{aligned}
a^{10} + 1 &= (10k \pm 3)^{10} + 1 \\
&\equiv 3^{10} + 1 \pmod{10} \\
&\equiv 0 \pmod{10}.
\end{aligned}
$$

- For $a = 10k \pm 4$, we obtain

$$
\begin{aligned}
a^{10} + 1 &= (10k \pm 4)^{10} + 1 \\
&\equiv 4^{10} + 1 \pmod{10} \\
&\equiv 7 \pmod{10}.
\end{aligned}
$$

- For $a = 10k \pm 5$, we obtain

$$
\begin{aligned}
a^{10} + 1 &= (10k \pm 5)^{10} + 1 \\
&\equiv 5^{10} + 1 \pmod{10} \\
&\equiv 6 \pmod{10}.
\end{aligned}
$$

Therefore, $a = 10k \pm 3$, where k is an integer.

Problem 25. Given m and n are two positive integers such that $m > n$. Prove that $a^{2^n} + 1$ divides $a^{2^m} - 1$ for all integers a.

Solution. Using the identity $x^2 - y^2 = (x - y)(x + y)$, we obtain

$$
\begin{aligned}
&(a - 1)(a + 1)\left(a^2 + 1\right) \cdots \left(a^{2^{m-1}} + 1\right) \\
&= \left(a^2 - 1\right)\left(a^2 + 1\right)\left(a^4 + 1\right) \cdots \left(a^{2^{m-1}} + 1\right) \\
&= \left(a^4 - 1\right)\left(a^4 + 1\right) \cdots \left(a^{2^{m-1}} + 1\right) \\
&= a^{2^m} - 1.
\end{aligned}
$$

Hence, $a^{2^k} + 1$ divides $a^{2^m} - 1$ for all $k \in \{0, 1, ..., m - 1\}$. Since $n < m$, it follows that $n \in \{0, 1, ..., m - 1\}$. Therefore, $a^{2^n} + 1$ divides $a^{2^m} - 1$ for all integers a.

Problem 26. For all positive integers n, prove the following statements:

1. $5^n - 2^n$ is divisible by 3;

2. $9^n + 16^n + 25^n - 2$ is divisible by 4;

3. $1^n + 5^n + 9^n + 13^n$ is divisible by 4;

4. $1000^n + 1001^n - (-1)^n - 1$ is divisible by 3.

Solution. For all positive integers n, prove the following statements:

1. $5^n - 2^n$ is divisible by 3;
   We have

$$5^n - 2^n \equiv (-1)^n - (-1)^n \pmod 3$$
$$\equiv 0 \pmod 3.$$

   Therefore, $5^n - 2^n$ is divisible by 3.

2. $9^n + 16^n + 25^n - 2$ is divisible by 4
   We have

$$9^n + 16^n + 25^n - 2 \equiv 1^n + 0^n + 1^n - 2 \pmod 4$$
$$\equiv 1 + 1 - 2 \pmod 4$$
$$\equiv 0 \pmod 4.$$

   Therefore $9^n + 16^n + 25^n - 2$ is divisible by 4.

3. $1^n + 5^n + 9^n + 13^n$ is divisible by 4
   We have

$$1^n + 5^n + 9^n + 13^n \equiv 1 + 1^n + 1^n + 1^n \pmod 4$$
$$\equiv 1 + 1 + 1 + 1 \pmod 4$$
$$\equiv 4 \pmod 4$$
$$\equiv 0 \pmod 4.$$

   Therefore, $1^n + 5^n + 9^n + 13^n$ is divisible by 4.

4. $1000^n + 1001^n - (-1)^n - 1$ is divisible by 3
   We have

$$1000^n + 1001^n - (-1)^n - 1 \equiv 1^n + (-1)^n - (-1)^n - 1 \pmod 3$$
$$\equiv 1 - 1 \pmod 3$$
$$\equiv 0 \pmod 3.$$

   Therefore, $1000^n + 1001^n - (-1)^n - 1$ is divisible by 3.

31

Problem 27. For all positive integers n, prove the following statements:

1. $9^n - 2^n$ is divisible by 7;

2. $4^{2n} + 5^{2n} + 1$ is divisible by 3;

3. $2^n - 5^n + 7^n - 10^n$ is divisible by 3;

4. $6^{2n+1} + 7^{2n+1} + 8^{2n+1} - 1$ is divisible by 5.

Solution. For all positive integers n, prove the following statements:

1. $9^n - 2^n$ is divisible by 7;
   We have

   $$9^n - 2^n \equiv 2^n - 2^n \pmod 7$$
   $$\equiv 0 \pmod 7.$$

   Therefore, $9^n - 2^n$ is divisible by 7.

2. $4^{2n} + 5^{2n} + 1$ is divisible by 3
   We have

   $$4^{2n} + 5^{2n} + 1 \equiv 1^{2n} + (-1)^{2n} + 1 \pmod 3$$
   $$\equiv 1 + 1 + 1 \pmod 3$$
   $$\equiv 3 \pmod 3$$
   $$\equiv 0 \pmod 3.$$

   Therefore, $4^{2n} + 5^{2n} + 1$ is divisible by 3.

3. $2^n - 5^n + 7^n - 10^n$ is divisible by 3
   We have

   $$2^n - 5^n + 7^n - 10^n \equiv (-1)^n - (-1)^n + 1^n - 1^n \pmod 3$$
   $$\equiv 0 \pmod 3.$$

   Therefore, $2^n - 5^n + 7^n - 10^n$ is divisible by 3.

4. $6^{2n+1} + 7^{2n+1} + 8^{2n+1} - 1$ is divisible by 5
   We have

   $$6^{2n+1} + 7^{2n+1} + 8^{2n+1} - 1$$

$$\equiv 1^{2n+1} + 2^{2n+1} + (-2)^{2n+1} - 1 \pmod 5$$
$$\equiv 1 + 2^{2n+1} - 2^{2n+1} - 1 \pmod 5$$
$$\equiv 0 \pmod 5.$$

Therefore, $6^{2n+1} + 7^{2n+1} + 8^{2n+1} - 1$ is divisible by 5.

Problem 28. Given a and b are two integers and n is a positive integers. Prove that $(a+b)^n \equiv nab^{n-1} + b^n \pmod{a^2}$.

Solution. From Newton's Binomial Theorem, we have

$$(a+b)^n = \sum_{k=0}^{n} a^{n-k} b^k C(n,k)$$
$$= \sum_{k=0}^{n-2} a^{n-k} b^k C(n,k) + ab^{n-1} C(n, n-1) + b^n$$
$$= \sum_{k=0}^{n-2} a^{n-k} b^k C(n,k) + nab^{n-1} + b^n$$
$$\equiv nab^{n-1} + b^n \pmod{a^2}.$$

Therefore, $(a+b)^n \equiv nab^{n-1} + b^n \pmod{a^2}$.

Remark 1. For the case $b = 1$, we obtain

$$(a+1)^n \equiv na + 1 \pmod{a^2}.$$

Problem 29. For all positive integers $n \geq 2$, prove the following statements:

1. $4^n - 3n - 1 \equiv 0 \pmod 9$;

2. $5^n - 4n - 1 \equiv 0 \pmod{16}$;

3. $6^n - 5n - 1 \equiv 0 \pmod{25}$;

4. $16 \times 4^n + 9 \times 5^n - 84n - 25 \equiv 0 \pmod{144}$.

Solution. For all positive integers $n \geq 2$, prove the following statements:

1. $4^n - 3n - 1 \equiv 0 \pmod 9$
   We have
   $$4^n = (3+1)^n$$

33

$$\equiv 3n + 1 \pmod 9.$$

Therefore, $4^n - 3n - 1 \equiv 0 \pmod 9$.

2. $5^n - 4n - 1 \equiv 0 \pmod{25}$
   We have

$$5^n = (4+1)^n$$
$$\equiv 4n + 1 \pmod{16}.$$

Therefore, $5^n - 4n - 1 \equiv 0 \pmod{16}$.

3. $6^n - 5n - 1 \equiv 0 \pmod{25}$
   We have

$$6^n = (5+1)^n$$
$$\equiv 5n + 1 \pmod{25}.$$

Therefore, $6^n - 5n - 1 \equiv 0 \pmod{25}$.

4. $16 \times 4^n + 9 \times 5^n - 84n - 25 \equiv 0 \pmod{144}$
   We have
$$4^n - 3n - 1 \equiv 0 \pmod 9$$
and
$$5^n - 4n - 1 \equiv 0 \pmod{16}.$$
It follows that

$$16 \times 4^n - 48n - 16 \equiv 0 \pmod{144} \qquad (1)$$

and
$$9 \times 5^n - 36n - 9 \equiv 0 \pmod{144}. \qquad (2)$$
Adding (1) and (2), we obtain

$$16 \times 4^n + 9 \times 5^n - 84n - 25 \equiv 0 \pmod{144}.$$

Therefore, $16 \times 4^n + 9 \times 5^n - 84n - 25 \equiv 0 \pmod{144}$.

Problem 30. Given a and b are two integers. For all positive integers $n \geq 2$, prove that

$$(a+b)^n \equiv \frac{n(n-1)}{2} a^2 b^{n-2} + nab^{n-1} + b^n \pmod{a^3}.$$

Solution. Using Newton's Binomial Theorem, we have

$$(a+b)^n = \sum_{k=0}^{n} a^{n-k} b^k C(n,k)$$

$$= \sum_{k=0}^{n-3} a^{n-k} b^k C(n,k) + a^2 b^{n-2} C(n,n-2)$$

$$+ ab^{n-1} C(n,n-1) + b^n$$

$$\equiv \frac{n!}{(n-2)!2!} a^2 b^{n-2} + nab^{n-1} + b^n \pmod{a^3}$$

$$\equiv \frac{n(n-1)}{2} a^2 b^{n-2} + nab^{n-1} + b^n \pmod{a^3}.$$

Therefore, $(a+b)^n \equiv \dfrac{n(n-1)}{2} a^2 b^{n-2} + nab^{n-1} + b^n \pmod{a^3}$.

Remark 2. For the case $b=1$, we obtain

$$(a+1)^n \equiv \frac{n(n-1)}{2} a^2 + na + 1 \pmod{a^3}.$$

Problem 31. For all positive integers $n \geq 3$, prove the following statements:

1. $2 \times 4^n - 9n^2 + 3n - 2$ is divisible by 27;
2. $5^n - 8n^2 + 4n - 1$ is divisible by 64;
3. $2 \times 6^n - 25n^2 + 15n - 2$ is divisible by 125.

Solution. For all positive integers $n \geq 3$, prove the following statements:

1. $2 \times 4^n - 9n^2 + 3n - 2$ is divisible by 27
   We have

$$4^n = (3+1)^n$$

$$\equiv \frac{n(n-1)}{2} \times 3^2 + 3n + 1 \pmod{3^3}$$

$$\equiv \frac{9}{2} n(n-1) + 3n + 1 \pmod{27}.$$

It follows that

$$2 \times 4^n \equiv 9n(n-1) + 6n + 2 \pmod{27}.$$

Then

$$2 \times 4^n - 9n^2 + 9n - 6n - 2 \equiv 0 \quad (\mathrm{mod}\ 27).$$

We obtain

$$2 \times 4^n - 9n^2 + 3n - 2 \equiv 0 \quad (\mathrm{mod}\ 27).$$

Therefore, $2 \times 4^n - 9n^2 + 3n - 2$ is divisible by 27.

2. $5^n - 8n^2 + 4n - 1$ is divisible by 64

We have

$$
\begin{aligned}
5^n &= (4+1)^n \\
&\equiv \frac{n(n-1)}{2} \times 4^2 + 4n + 1 \quad (\mathrm{mod}\ 4^3) \\
&\equiv 8n(n-1) + 4n + 1 \quad (\mathrm{mod}\ 64) \\
&\equiv 8n^2 - 8n + 4n + 1 \quad (\mathrm{mod}\ 64) \\
&\equiv 8n^2 - 4n + 1 \quad (\mathrm{mod}\ 64).
\end{aligned}
$$

Hence, $5^n - 8n^2 + 4n - 1 \equiv 0 \ (\mathrm{mod}\ 64)$.

Therefore, $5^n - 8n^2 + 4n - 1$ is divisible by 64.

3. $2 \times 6^n - 25n^2 + 15n - 2$ is divisible by 125

We have

$$
\begin{aligned}
6^n &= (5+1)^n \\
&\equiv \frac{n(n-1)}{2} \times 5^2 + 5n + 1 \quad (\mathrm{mod}\ 5^2) \\
&\equiv \frac{25}{2} n(n-1) + 5n + 1 \quad (\mathrm{mod}\ 125).
\end{aligned}
$$

We obtain

$$2 \times 6^n \equiv 25n(n-1) + 10n + 2 \quad (\mathrm{mod}\ 125).$$

It follows that

$$2 \times 6^n \equiv 25n^2 - 25n + 10n + 2 \quad (\mathrm{mod}\ 125).$$

Hence,

$$2 \times 6^n \equiv 25n^2 - 15n + 2 \quad (\mathrm{mod}\ 125).$$

It implies that

$$2 \times 6^n - 25n^2 + 15n - 2 \equiv 0 \quad (\mathrm{mod}\ 125).$$

Therefore, $2 \times 6^n - 25n^2 + 15n - 2$ is divisible by 125.

Problem 32. Prove that $7^{19} + 8^{19} + 9^{19} + 10^{19} + 11^{19} + 12^{19}$ is divisible by 19.

Solution. Since 19 is an odd number, we obtain

$$7^{19} + 8^{19} + 9^{19} + 10^{19} + 11^{19} + 12^{19}$$
$$= \left(7^{19} + 12^{19}\right) + \left(8^{19} + 11^{19}\right) + \left(9^{19} + 10^{19}\right)$$
$$\equiv 7 + 12 + 8 + 11 + 9 + 10 \pmod{19}$$
$$\equiv 19 + 19 + 19 \pmod{19}$$
$$\equiv 0 \pmod{19}.$$

Therefore, $7^{19} + 8^{19} + 9^{19} + 10^{19} + 11^{19} + 12^{19}$ is divisible by 19.

Problem 33. Given x and y are two positive integers and p is an odd prime. Prove that

$$(x + y)^{p-1} \equiv \sum_{k=1}^{p-1} x^{p-1-k} y^k \pmod{p}.$$

Solution. From Newton's Binomial Theorem, we have

$$(x + y)^{p-1} \equiv \sum_{k=1}^{p-1} (-1)^k C(p-1, k) x^{p-1-k} y^k.$$

Hence, to prove the given statement, it is sufficient to prove that

$$(-1)^k C(p-1, k) \equiv 1 \pmod{p}.$$

We have $C(p-1, k) = \dfrac{(p-1)!}{k!(p-1-k)!}.$

For all integers k such that $0 \le k \le p-1$, we have

$$p - 1 - k \equiv (-1)(k+1) \pmod{p}.$$

It follows that

$$(p-1-k)! \equiv (-1)^{p-k-1}(k+1)\ldots(p-1) \pmod{p}.$$

Multiply both sides of the congruence by $(-1)^k k!$, we obtain

$$(-1)^k k! (p-1-k)! \equiv (-1)^{p-1}(p-1)!$$
$$\equiv (p-1)! \pmod{p}.$$

37

Moreover, $\gcd\left(k!\,(p-1-k)!,p\right) = 1$ for all $0 \le k < p$. It implies that

$$\frac{(-1)^k\,(p-1)!}{k!\,(p-1-k)!} \equiv 1 \quad (\bmod\ p)$$

or

$$(-1)^k C\,(p-1,k) \equiv 1 \quad (\bmod\ p).$$

Therefore, $(x+y)^{p-1} \equiv \displaystyle\sum_{k=1}^{p-1} x^{p-1-k} y^k \quad (\bmod\ p)$.

Problem 34. Given $3n$ integers $x_1, x_2, ..., x_{3n}$ which are not divisible by 3. Prove that $x_1^2 + x_2^2 + ... + x_{3n}^2$ is divisible by 3.

Solution. For all integers x which is not divisible by 3, we have $x \equiv 1$ or $2 \pmod 3$. It follows that $x^2 \equiv 1 \pmod 3$. We obtain

$$x_1^2 + x_2^2 + ... + x_{3n}^2 \equiv \underbrace{1 + 1 + ... + 1}_{3n} \quad (\bmod\ 3)$$

$$\equiv 3n \quad (\bmod\ 3)$$

$$\equiv 0 \quad (\bmod\ 3).$$

Therefore, $x_1^2 + x_2^2 + ... + x_{3n}^2$ is divisible by 3.

Problem 35. Given four integers a, b, c, d and two positive integers m and $n \ge 2$ such that $a \equiv b \pmod m$ and $c \equiv d \pmod n$. Prove that $ac + bd \equiv ad + bc \pmod{mn}$.

Solution. Since $a \equiv b \pmod m$ and $c \equiv d \pmod n$, it follows that $a - b = mk$ and $c - d = nl$ for some integers k and l. We obtain

$$(a-b)\,(c-d) = mnkl.$$

Then $ac - ad - bc + bd = mnkl$.
It follows that $(ac+bd) - (ad+bc) = mnkl$.
Therefore, $ac + bd \equiv ad + bc \pmod{mn}$.

Problem 36. Given two positive integers a and b such that $a > b$. Prove that

$$(a-1)\,(a-2)...\,(a-b) \equiv (-1)^b b! \quad (\bmod\ a).$$

Solution. For all positive integers k, we have $a - k \equiv -k \pmod{a}$. Hence, $a - 1 \equiv -1 \pmod{a}, a - 2 \equiv -2 \pmod{a}, ..., a - b \equiv -b \pmod{a}$. It follows that

$$(a - 1)(a - 2)...(a - b) \equiv (-1)(-2)...(-b) \pmod{a}$$
$$\equiv (-1)^b b! \pmod{a}.$$

Therefore, $(a - 1)(a - 2)...(a - b) \equiv (-1)^b b! \pmod{a}$.

Problem 37. Let p be a prime number. Prove that $p \mid C(p, k)$ for all positive integer k such that $1 \le k < p$.

Solution. We have

$$C(p, k) = \frac{p!}{k!(p - k)!}$$
$$= \frac{p(p - 1)...(p - k + 1)(p - k)!}{k!(p - k)!}$$
$$= \frac{p(p - 1)...(p - k + 1)}{k!}.$$

Then $k!C(p, k) = p(p - 1)...(p - k + 1)$.
It turns out that $p \mid k!C(p, k)$.
Since p is a prime number and $1 \le k < p$, we obtain $\gcd(p, k!) = 1$. Hence, $p \mid C(p, k)$.

Problem 38. Let p be a prime number. Prove that $p \mid C(2p, p) - p$.

Solution. We have

$$C(2p, p) = \frac{(2p)!}{p!p!}$$
$$= \frac{(2p)(2p - 1)...(p - 1)p!}{p!p!}$$
$$= \frac{(2p)(2p - 1)...(p - 1)}{p!}$$
$$= \frac{2(2p - 1)(2p - 2)...(p - 1)}{(p - 1)!}.$$

We know that $2p - 1 \equiv p - 1 \pmod{p}, 2p - 2 \equiv p - 2 \pmod{p}, ..., p - 1 \equiv 1 \pmod{p}$. Hence,

$$(2p - 1)(2p - 2)...(p - 1) \equiv (p - 1)!.$$

39

Since p is a prime number, it follows that $\gcd((p-1)!, p) = 1$. We obtain

$$\frac{(2p-1)(2p-2)\ldots(p-1)}{(p-1)!} \equiv 1 \pmod{p}.$$

It implies that $C(2p, p) \equiv 2 \pmod{p}$.
Therefore, $p \mid C(2p, p) - p$.

Problem 39. Prove that perfect squares are of the form $4k$ or $4k+1$, where k is a nonnegative integer.

Solution. Let n be a perfect square. Then $n = a^2$ for some integers a. For all integers a, we obtain $a = 2l$ or $2l+1$ for some integers l.

- For $a = 2l$, it follows that $n = (2l)^2 = 4l^2 = 4k$, where $k = l^2$ is a nonnegative integer.

- For $a = 2l+1$, it implies that

$$\begin{aligned} n &= (2l+1)^2 \\ &= 4l^2 + 4l + 1 \\ &= 4(l^2 + l) + 1 \\ &= 4k + 1 \end{aligned}$$

, where $k = l^2 + l$ is a nonnegative integer.

Problem 40. For all positive integers n, prove that

$$n^4 + 2n^3 + 2n^2 + 2n + 1$$

is never a perfect square.

Solution. We shall prove the given statement by contradiction. Suppose that there exists a positive integer n such that $n^4 + 2n^3 + 2n^2 + 2n + 1$ is a perfect square. We have

$$\begin{aligned} n^4 + 2n^3 + 2n^2 + 2n + 1 &= n^2(n^2 + 2n + 1) + (n^2 + 2n + 1) \\ &= (n^2 + 2n + 1)(n^2 + 1) \\ &= (n+1)^2(n^2 + 1). \end{aligned}$$

Since $n^4 + 2n^3 + 2n^2 + 2n + 1$ is a perfect square, we obtain $n^2 + 1$ is a perfect square. Let $n^2 + 1 = t^2$, where t is a positive integer. Then $(n - t)(n + t) = -1$. Since $n + t > n - t$, we obtain

$$\begin{cases} n - t = -1 \\ n + t = 1 \end{cases}.$$

Adding both sides of the two equations, we obtain $2n = 0$. Then $n = 0$, a contradiction.

Therefore, $n^4 + 2n^3 + 2n^2 + 2n + 1$ is never a perfect square.

Problem 41. Given that p_n is the nth prime number. That is, $p_1 = 2, p_2 = 3, p_3 = 5, ...$ Prove that $H_n = p_1 p_2 p_3 ... p_n + 1$ is never a perfect square for all positive integers n.

Solution. For the case $n = 1$, we obtain $H_1 = 2 + 1 = 3$, not a perfect square. For all $n \geq 2$, we have

$$H_n = p_1 p_2 p_3 ... p_n + 1$$
$$= 2(p_2 p_3 ... p_n) + 1.$$

Since $p_2, p_3, ..., p_n$ are odd numbers, it follows that $p_2 p_3 ... p_n$ is also an odd number. It turns out that $p_2 p_3 ... p_n = 2k + 1$ for some nonnegative integers k. Hence,

$$H_n = 2(2k + 1) + 1$$
$$= 4k + 2 + 1$$
$$= 4k + 3$$
$$\equiv 3 \pmod{4}.$$

Then H_n is never a perfect square for all $n \geq 2$.

Therefore, H_n is never a perfect square for all $n \geq 1$.

Problem 42. Find all positive integers n such that

$$n^4 + 4n^3 + 7n^2 + 12n + 12$$

is a perfect square.

Solution. We have

$$n^4 + 4n^3 + 7n^2 + 12n + 12$$
$$= n^2(n^2 + 4n + 4) + 3(n^2 + 4n + 4)$$

41

$$= \left(n^2 + 4n + 4\right)\left(n^2 + 3\right)$$
$$= (n+2)^2 \left(n^2 + 3\right).$$

Since $n^4 + 4n^3 + 7n^2 + 12n + 12$ is a perfect square, we obtain $n^2 + 3$ is a perfect square. Let $n^2 + 3 = t^2$, where t is a positive integer. It follows that

$$(n - t)(n + t) = -3.$$

Since $n + t > n - t$, we obtain $\begin{cases} n + t = 3 \\ n - t = -1 \end{cases}$.

Adding both sides of the two equations, we obtain $2n = 2$. It implies that $n = 1$.
For $n = 1$, we obtain

$$n^4 + 4n^3 + 7n^2 + 12n + 12$$
$$= 1 + 4 + 7 + 12 + 12$$
$$= 36, \text{ is a perfect square.}$$

Therefore, $n = 1$.

A Fibonacci sequence $\{F_n\}$ is a sequence such that $F_1 = F_2 = 1$ and $F_{n+2} = F_{n+1} + F_n$ for all $n \geq 1$. F_n is said to be the nth Fibonacci number.

Problem 43. (Cassini's Formula)
Let $\{F_n\}$ be a Fibonacci sequence. For all positive integers $n \geq 2$, prove that

$$F_{n+1}F_{n-1} - F_n^2 = (-1)^n.$$

Solution. We will prove the given statement by induction. Observe that

$$F_3 F_1 - F_2^2 = (2)(1) - 1^2 = 1 = (-1)^2.$$

Hence, the given statement holds for $n = 2$. Now, let us suppose that $F_{m+1}F_{m-1} - F_m^2 = (-1)^m$ for some integers $m \geq 2$. We shall show that

$$F_{m+2}F_m - F_{m+1}^2 = (-1)^{m+1}.$$

We have

$$F_{m+2}F_m - F_{m+1}^2 = (F_{m+1} + F_m)(F_{m+1} - F_{m-1}) - F_{m+1}^2$$
$$= F_{m+1}^2 - F_{m+1}F_{m-1} + F_m(F_{m+1} - F_{m-1}) - F_{m+1}^2$$

$$= -F_{m+1}F_{m-1} + F_m^2$$
$$= -\left(F_{m+1}F_{m-1} - F_m^2\right)$$
$$= -(-1)^m$$
$$= (-1)^{m+1}.$$

Therefore, $F_{n+1}F_{n-1} - F_n^2 = (-1)^n$.

Problem 44. Prove that any two consecutive Fibonacci numbers are relatively prime.

Solution. To prove the given statement, we will prove that

$$\gcd(F_m, F_{m+1}) = 1.$$

Let d be the greatest common divisor of F_m and F_{m+1}. Suppose that $d \geq 1$. We obtain $d \mid F_m$ and $d \mid F_{m+1}$. It follows that

$$d \mid F_{m+1}F_{m-1} - F_m^2 = (-1)^m.$$

It turns out that $d = 1$.
Therefore, any two consecutive Fibonacci numbers are relatively prime.

Fermat's numbers are numbers of the form $f_n = 2^{2^n} + 1$, where $n \geq 0$.

Problem 45. For all $n \geq 1$, prove that

$$f_0 f_1 \ldots f_{n-1} = f_n - 2.$$

Solution. We shall prove the given statement by induction.
For $n = 1$, the left-hand side of the identity equals $f_0 = 2^{2^0} + 1 = 3$.
The right-hand side of the identity equals $f_1 - 2 = 2^{2^1} + 1 - 2 = 3$.
It turns out that the given statement holds for the case $n = 1$.
Suppose that the statement holds for some $n = k$, where $k \geq 1$.
We will show that the statement holds for $n = k + 1$.
From the induction hypothesis, we have

$$f_0 f_1 \ldots f_{k-1} = f_k - 2.$$

It follows that

$$f_0 f_1 \ldots f_{k-1} f_k = (f_0 f_1 \ldots f_{k-1}) f_k$$

$$= (f_k - 2) f_k$$
$$= \left(2^{2^k} + 1 - 2\right)\left(2^{2^k} + 1\right)$$
$$= \left(2^{2^k} - 1\right)\left(2^{2^k} + 1\right)$$
$$= \left(2^{2^k}\right)^2 - 1$$
$$= 2^{2^{k+1}} - 1$$
$$= f_{k+1} - 2.$$

Therefore, $f_0 f_1 ... f_{n-1} = f_n - 2$ for all $n \geq 1$.

Problem 46. Given two distinct nonnegative integers m and n. Prove that f_m and f_n are relatively prime.

Solution. Without loss of generality, suppose that $m < n$. It follows that

$$f_0 f_1 ... f_m ... f_{n-1} = f_n - 2.$$

Then $f_n - f_0 f_1 ... f_m ... f_{n-1} = 2$.
Let $d = \gcd(f_m, f_n)$. It implies that $d \mid f_m$ and $d \mid f_n$.
Hence, $d \mid f_n - f_0 f_1 ... f_m ... f_{n-1} = 2$.
We obtain $d = 1$ or 2.
However, f_n and f_m are odd numbers. Hence, $d = 1$.
Therefore, f_m and f_n are relatively prime.

Problem 47. For all prime numbers p, prove that p divides

$$5^p + 3^p + 2^p - 10.$$

Solution. For all prime numbers p, using Fermat's little theorem, we have $5^p \equiv 5 \pmod{p}$, $3^p \equiv 3 \pmod{p}$ and $2^p \equiv 2 \pmod{p}$. It follows that

$$5^p + 3^p + 2^p - 10 \equiv 5 + 3 + 2 - 10 \pmod{p}$$
$$\equiv 0 \pmod{p}.$$

Therefore, p divides $5^p + 3^p + 2^p - 10$.

Problem 48. Given three integers a, b, c and a prime number p such that $\gcd(a, p) = \gcd(b, p) = \gcd(c, p) = 1$ and $p \mid a + b + c$. Prove that p divides $ab^{p-1} + bc^{p-1} + ca^{p-1}$.

44

Solution. Since p is a prime number and $\gcd(a,p) = \gcd(b,p) = \gcd(c,p) = 1$, using Fermat's little theorem, we have $a^{p-1} \equiv 1 \pmod{p}, b^{p-1} \equiv 1 \pmod{p}$ and $c^{p-1} \equiv 1 \pmod{p}$. Hence,

$$ab^{p-1} + bc^{p-1} + ca^{p-1} \equiv a + b + c \pmod{p}.$$

By knowing that $p \mid a + b + c$, we obtain $a + b + c \equiv 0 \pmod{p}$. It turns out that $ab^{p-1} + bc^{p-1} + ca^{p-1} \equiv 0 \pmod{p}$. Therefore, p divides $ab^{p-1} + bc^{p-1} + ca^{p-1}$.

Problem 49. For all integers a, prove that 30 divides $a^{10} - a^2$.

Solution. Since 2 is a prime number, using Fermat's little theorem, we have

$$a^2 \equiv a \pmod{2}.$$

We obtain

$$\begin{aligned} a^{10} &= \left(a^2\right)^5 \\ &\equiv a^5 \pmod{2} \\ &\equiv \left(a^2\right)^2 \times a \pmod{2} \\ &\equiv a^2 \times a \pmod{2} \\ &\equiv a \times a \pmod{2} \\ &\equiv a^2 \pmod{2}. \end{aligned} \tag{1}$$

Moreover, we have $a^3 \equiv a \pmod{3}$. It follows that

$$\begin{aligned} a^{10} &= \left(a^3\right)^3 \times a \\ &\equiv a^3 \times a \pmod{3} \\ &\equiv a \times a \pmod{3} \\ &\equiv a^2 \pmod{3}. \end{aligned} \tag{2}$$

Using Fermat's little theorem again, we have $a^5 \equiv a \pmod{5}$. Hence,

$$\begin{aligned} a^{10} &= \left(a^5\right)^2 \\ &\equiv a^2 \pmod{5}. \end{aligned} \tag{3}$$

From (1), (2) and (3), since 2, 3 and 5 are pairwise relatively prime, we obtain $a^{10} \equiv a^2 \pmod{(2 \times 3 \times 5)}$. That is, $a^{10} \equiv a^2 \pmod{30}$. Therefore, 30 divides $a^{10} - a^2$.

Problem 50. For all integers a, prove that 105 divides $a^{13} - a$.

Solution. Observe that $105 = 3 \times 5 \times 7$. Hence, to prove that 105 divides $a^{13} - a$, it is sufficient to prove that 3, 5 and 7 divide $a^{13} - a$. Since 3, 5 and 7 are prime numbers, using Fermat's little theorem, we have $a^3 \equiv a \pmod 4$, $a^5 \equiv a \pmod 5$ and $a^7 \equiv a \pmod 7$. We have

$$a^{13} = \left(a^3\right)^4 \times a$$
$$\equiv a^4 \times a \pmod 3$$
$$\equiv a^3 \times a^2 \pmod 3$$
$$\equiv a \times a^2 \pmod 3$$
$$\equiv a^3 \pmod 3$$
$$\equiv a \pmod 3. \tag{1}$$

On the other hand,

$$a^{13} = \left(a^5\right)^2 \times a^3$$
$$\equiv a^2 \times a^3 \pmod 5$$
$$\equiv a^5 \pmod 5$$
$$\equiv a \pmod 5. \tag{2}$$

Moreover,

$$a^{13} = a^7 \times a^6$$
$$\equiv a \times a^6 \pmod 7$$
$$\equiv a^7 \pmod 7$$
$$\equiv a \pmod 7. \tag{3}$$

From (1), (2) and (3), since 3, 5 and 7 are pairwise relatively prime, we obtain

$$a^{13} \equiv a \pmod{3 \times 5 \times 7}.$$

Then $a^{13} \equiv a \pmod{105}$.
Therefore, 105 divides $a^{13} - a$.

Problem 51. For all integers a, prove that 119 divides $a^{100} - a^4$.

Solution. Since 7 and 17 are prime numbers, using Fermat's little theorem, we have $a^7 \equiv a \pmod 7$ and $a^{17} \equiv a \pmod{17}$. It follows that

$$a^{100} = \left(a^7\right)^{14} \times a^2$$

46

$$\equiv a^{14} \times a^2 \pmod 7$$
$$\equiv \left(a^7\right)^2 \times a^2 \pmod 7$$
$$\equiv a^2 \times a^2 \pmod 7$$
$$\equiv a^4 \pmod 7. \tag{1}$$

Furthermore,

$$a^{100} = \left(a^{17}\right)^5 \times a^{15}$$
$$\equiv a^5 \times a^{15} \pmod{17}$$
$$\equiv a^{17} \times a^3 \pmod{17}$$
$$\equiv a \times a^3 \pmod{17}$$
$$\equiv a^4 \pmod{17}. \tag{2}$$

Since $\gcd(7, 17) = 1$, from (1) and (2), we obtain

$$a^{100} \equiv a^4 \pmod{7 \times 17}.$$

Then $a^{100} \equiv a^4 \pmod{119}$.
Therefore, 119 divides $a^{100} - a^4$.

Problem 52. Given that a is an integer such that $\gcd(a, 15) = 1$. Prove that 15 divides $a^{12} - 1$.

Solution. Since $\gcd(a, 15) = 1$, it follows that $\gcd(a, 3) = \gcd(a, 5) = 1$. Using Fermat's little theorem, we obtain

$$a^2 \equiv 1 \pmod 3$$

and

$$a^4 \equiv 1 \pmod 5.$$

Hence,

$$a^{12} \equiv 1 \pmod 3$$

and

$$a^{12} \equiv 1 \pmod 5.$$

Since $\gcd(3, 5) = 1$, it implies that $a^{12} \equiv 1 \pmod{15}$.
Therefore, 15 divides $a^{12} - 1$.

Problem 53. Given that a is an integer such that $\gcd(a, 14) = 1$. Prove that 28 divides $a^{12} - 1$.

Solution. Since $\gcd(a, 14) = 1$, it follows that

$$\gcd(a, 2) = \gcd(a, 7) = 1.$$

Using Fermat's little theorem, we obtain

$$a \equiv 1 \pmod{2}$$

and

$$a^6 \equiv 1 \pmod{7}.$$

- From $a \equiv 1 \pmod 2$, we obtain $a = 2k + 1$ for some integers k. Then

$$\begin{aligned} a^2 &= (2k + 1)^2 \\ &= 4k^2 + 4k + 1 \\ &\equiv 1 \pmod 4. \end{aligned}$$

It follows that

$$a^{12} \equiv 1. \pmod 4 \tag{1}$$

- From $a^6 \equiv 1 \pmod 7$, we obtain

$$a^{12} \equiv 1 \pmod 7. \tag{2}$$

Since $\gcd(4, 7) = 1$, from (1) and (2), we obtain

$$a^{12} \equiv 1 \pmod{24}.$$

Therefore, 28 divides $a^{12} - 1$.

Problem 54. Given a is an integer and p, q are two distinct prime numbers such that $\gcd(a, pq) = 1$. Prove that

$$p^2 a^{2(q-1)} + q^2 a^{2(p-1)} \equiv p^2 + q^2 \pmod{pq}.$$

Solution. Since a is an integer and p, q are two distinct prime numbers such that $\gcd(a, pq) = 1$, we obtain $\gcd(a, p) = \gcd(a, q) = 1$. Using Fermat's little theorem, we obtain

$$a^{p-1} \equiv 1 \pmod p$$

and

$$a^{q-1} \equiv 1 \pmod q.$$

It follows that
$$qa^{p-1} \equiv q \pmod{pq} \tag{1}$$

and
$$pa^{q-1} \equiv p \pmod{pq}. \tag{2}$$

Adding (1) and (2), we obtain
$$pa^{q-1} + qa^{p-1} \equiv p + q \pmod{pq}.$$

Squaring both sides of the above modulo congruence, we obtain
$$p^2 a^{2(q-1)} + 2pqa^{p+q-2} + q^2 a^{2(p-1)} \equiv p^2 + 2pq + q^2 \pmod{pq}.$$

Then $p^2 a^{2(q-1)} + q^2 a^{2(p-1)} \equiv p^2 + q^2 \pmod{pq}$.
Therefore, $p^2 a^{2(q-1)} + q^2 a^{2(p-1)} \equiv p^2 + q^2 \pmod{pq}$.

Problem 55. Given a is an integer such that $\gcd(a, 30) = 1$. Prove that $a^4 - 1$ is divisible by 60.

Solution. Since $\gcd(a, 30) = 1$, we obtain $\gcd(a, 2) = \gcd(a, 3) = \gcd(a, 5) = 1$.

- For $\gcd(a, 2) = 1$, we obtain $a = 2k + 1$ for some integers k. It follows that
$$\begin{aligned} a^2 &= (2k+1)^2 \\ &= 4k^2 + 4k + 1 \\ &\equiv 1 \pmod{4}. \end{aligned}$$

  Then
$$a^4 \equiv 1 \pmod{4}. \tag{1}$$

- For $\gcd(a, 3) = 1$, using Fermat's little theorem, we obtain $a^2 \equiv 1 \pmod{3}$.
  Then
$$a^4 \equiv 1 \pmod{3}. \tag{2}$$

- For $\gcd(a, 5) = 1$, using Fermat's little theorem, we obtain
$$a^4 \equiv 1 \pmod{5}. \tag{3}$$

From (1), (2) and (3), since $3, 4$ and 5 are pairwise relatively prime, we obtain

$$a^4 \equiv 1 \pmod{4 \times 3 \times 5}.$$

It implies that $a^4 \equiv 1 \pmod 6)0$.
Therefore, $a^4 - 1$ is divisible by 60.

Problem 56. Given p and q are two prime numbers such that $p > q$. Prove that q divides $3^p - 2^q - 3^{p-q+1} + 2$.

Solution. We know that p and q are two prime numbers such that $p > q$. Then $p = q + k$ for some positive integers k. We obtain

$$
\begin{aligned}
3^p - 2^q &= 3^{q+k} - 2^q \\
&= (2+1)^q \times 3^k - 2^q \\
&= 3^k \sum_{i=0}^{q} C(q,i) 2^{q-i} - 2^q \\
&= 3^k \left(2^q + 1 + \sum_{i=1}^{q-1} C(q,i) 2^{q-i} \right) - 2^q \\
&\equiv 3^k (2^q + 1) - 2^q \pmod q.
\end{aligned}
$$

From Fermat's little theorem, we have $2^q \equiv 2 \pmod q$. It implies that

$$
\begin{aligned}
3^p - 2^q &\equiv 3^k (2+1) - 2 \pmod q \\
&\equiv 3^{k+1} - 2 \pmod q \\
&\equiv 3^{p-q+1} - 2 \pmod q.
\end{aligned}
$$

Then $3^p - 2^q - 3^{p-q+1} + 2 \equiv 0 \pmod q$.
Therefore, $q \mid 3^p - 2^q - 3^{p-q+1} + 2$.

Problem 57. Suppose that p and q are two distinct prime numbers such that $a^p \equiv a \pmod q$ and $a^q \equiv a \pmod p$. Prove that

$$a^{pq} \equiv a \pmod{pq}.$$

Solution. From Fermat's little theorem, we have

$$a^{pq} = (a^q)^p \equiv a^q \pmod p$$

and

$$a^q \equiv a \pmod p.$$

It follows that $a^{pq} = a$ (mod p).

Similarly, $a^{pq} = a$ (mod q).

Consequently, $a^{pq} = a$ (mod pq) since $\gcd(p, q) = 1$.

Therefore, $a^{pq} \equiv a$ (mod pq).

Problem 58. Given two distinct prime numbers p and q. Prove that

$$p^{q-1} + q^{p-1} \equiv 1 \pmod{pq}.$$

Solution. Since p and q are two distinct prime numbers, using Fermat's little theorem, we obtain $p^{q-1} \equiv 1$ (mod q). It is obvious to see that $q^{p-1} \equiv 0$ (mod q).

It follows that

$$p^{q-1} + q^{p-1} \equiv 1 \pmod{q}. \tag{1}$$

Similarly,

$$p^{q-1} + q^{p-1} \equiv 1 \pmod{p}. \tag{2}$$

Since $\gcd(p, q) = 1$, from (1) and (2), we obtain

$$p^{q-1} + q^{p-1} \equiv 1 \pmod{pq}.$$

Therefore, $p^{q-1} + q^{p-1} \equiv 1$ (mod pq).

Problem 59. Let a and b be positive integers. Let p be a prime number such that $\gcd(a, p) = \gcd(b, p) = 1$. Prove the following statements:

1. If $a^p \equiv b^p$ (mod p), then $a \equiv b$ (mod p).

2. If $a^p \equiv b^p$ (mod p), then $a^p \equiv b^p$ (mod p^2).

Solution. Prove the following statements:

1. If $a^p \equiv b^p$ (mod p), then $a \equiv b$ (mod p).
   From Fermat's little theorem, we have $a^p \equiv a$ (mod p) and $b^p \equiv b$ (mod p). Using the fact that $a^p \equiv b^p$ (mod p) and transitivity of congruence, it follows that $a \equiv b$ (mod p). Therefore, $a \equiv b$ (mod p).

2. If $a^p \equiv b^p$ (mod p), then $a^p \equiv b^p$ (mod p^2).
   Since $a^p \equiv b^p$ (mod p), from the above proof, we have $a \equiv b$ (mod p). Then there exists an integer k such that $a = b + pk$. Observe that

$$a^p - b^p = (b + pk)^p - b^p$$

51

$$= b^p + \sum_{i=1}^{p} \binom{p}{i} b^{p-i}(pk)^i - b^p$$

$$= \sum_{i=1}^{p} \frac{p!}{i!\,(p-i)!} b^{p-i}(pk)^i.$$

For all $i \geq 2$, it is obvious to see that each terms has p^2. That is,

$$p^2 \Big| \sum_{i=2}^{p} \frac{p!}{i!\,(p-i)!} b^{p-i}(pk)^i.$$

For the case $i = 1$, we have $\dfrac{p!}{1!\,(p-1)!} b^{p-1}pk = b^{p-1}kp^2$ which is divisible by p^2. It implies that

$$p^2 \Big| \sum_{i=1}^{p} \frac{p!}{i!\,(p-i)!} b^{p-i}(pk)^i.$$

Therefore, $a^p \equiv b^p \pmod{p^2}$.

Problem 60. Let a and b be two positive integers and p be a prime number. Prove that $(a+b)^p \equiv a^p + b^p \pmod p$.

Solution. Method I:
Since a and b are two positive integers and p is a prime number, then

$$a^p \equiv a \pmod p$$

and

$$b^p \equiv b \pmod p.$$

It follows that

$$a^p + b^p \equiv a + b \pmod p. \tag{1}$$

Using Fermat's little theorem, we obtain

$$(a+b)^p \equiv a + b \pmod p. \tag{2}$$

From (1) and (2), we obtain $(a+b)^p \equiv a^p + b^p \pmod p$.
Method II:
Using Newton's Binomial Theorem, we have

$$(a+b)^p = \sum_{k=0}^{p} C(p,k)\, a^{p-k} b^k$$

$$= C(p,0)a^p + \sum_{k=1}^{p-1} C(p,k)a^{p-k}b^k + C(p,p)b^p$$

$$= a^p + b^p + \sum_{k=1}^{p-1} C(p,k)a^{p-k}b^k.$$

From Problem 37, we know that $p \mid C(p,k)$ for all $1 \le k \le p-1$.

Then $p \mid \sum_{k=1}^{p-1} C(p,k)a^{p-k}b^k$ (mod p).

Therefore, $(a+b)^p \equiv a^p + b^p$ (mod p).

Problem 61. Let p be a prime number and x and y be two integers. Suppose that $x \equiv y$ (mod $p-1$). Prove that $a^x \equiv a^y$ (mod p) for all positive integers a such that $\gcd(a,p) = 1$.

Solution. Since $x \equiv y$ (mod $p-1$), we obtain $x = y + (p-1)k$ for some integers k. It follows that

$$a^x = a^{y+(p-1)k}$$

$$= a^y \times \left(a^{p-1}\right)^k.$$

By knowing that p is a prime number and $\gcd(a,p) = 1$, applying Fermat's little theorem, we obtain $a^{p-1} \equiv 1$ (mod p). It follows that

$$a^x \equiv a^y \times 1^k \quad \text{(mod } p\text{)}.$$

Therefore, $a^x \equiv a^y$ (mod p).

Problem 62. Given p is a prime number and a is an integer such that $\gcd(a,p) = 1$. Prove that a^{p-2} is the inverse of a modulo p.

Solution. Since p is a prime number and a is an integer such that $\gcd(a,p) = 1$, applying Fermat's little theorem, we have

$$a^{p-1} \equiv 1 \quad \text{(mod } p\text{)}.$$

It implies that

$$a \times a^{p-2} \equiv 1 \quad \text{(mod } p\text{)}.$$

Therefore, a^{p-2} is the inverse of a modulo p.

Problem 63. Let $p > 3$ be a prime number. Prove that

$$a^p \equiv a \quad \text{(mod } 6p\text{)}.$$

Solution. Since $p > 3$ is a prime number, using Fermat's little theorem, we obtain $a^p \equiv a \pmod p$. Hence, to prove the statement, it is sufficient to prove that $a^p \equiv a \pmod 6$.
Observe that

$$a^p - a = a\left(a^{p-1} - 1\right)$$
$$= a\left(a - 1\right)\left(a^{p-2} + a^{p-3} + \dots + 1\right).$$

Since $a - 1$ and a are two consecutive integers, we obtain

$$2 \mid a(a - 1).$$

Hence,

$$2 \mid a^p - a. \tag{1}$$

On the other hand, for all integers a, we obtain $a = 3k - 1, 3k$ or $3k + 1$ for some integers k.

- For $a = 3k \pm 1$, we obtain

$$a^p - a = a\left[(3k \pm 1)^{p-1} - 1\right]$$
$$\equiv a\left[(-1)^{p-1} - 1\right] \pmod 3$$
$$\equiv a\left(1 - 1\right) \pmod 3$$
$$\equiv 0 \pmod 3.$$

- For $a = 3k$, we obtain $a^p - a \equiv 0 \pmod 3$.

Hence,

$$3 \mid a^p - a. \tag{2}$$

From (1) and (2), since $\gcd(2, 3) = 1$, we obtain $a^p - a$ is divisible by 6.
Therefore, $a^p \equiv a \pmod{6p}$.

Problem 64. Suppose that p and q are two distinct odd prime numbers such that $p - 1 \mid q - 1$. If $\gcd(a, pq) = 1$, prove that

$$a^{q-1} \equiv 1 \pmod{pq}.$$

Solution. Since $\gcd(a, pq) = 1$ and p, q are two distinct odd prime numbers, it follows that $\gcd(a, p) = \gcd(a, q) = 1$.
Using Fermat's little theorem, we obtain

$$a^{p-1} \equiv 1 \pmod p$$

and
$$a^{q-1} \equiv 1 \pmod{q}.$$

Since $p - 1 \mid q - 1$, then there exist an integer k such that

$$q - 1 = k(p - 1).$$

Hence, $a^{p-1} \equiv 1 \pmod{p}$ implies that

$$a^{k(p-1)} \equiv 1 \pmod{p}.$$

Consequently, $a^{q-1} \equiv 1 \pmod{p}$.
As a result, $p \mid a^{q-1} - 1$ and $q \mid a^{q-1} - 1$.
Using the fact that $\gcd(p, q) = 1$, it follows that $pq \mid a^{q-1} - 1$.
Therefore, $a^{q-1} \equiv 1 \pmod{pq}$.

Problem 65. Let n be a positive integer. Suppose that $3^n - 2^n$ is a power of prime number. Prove that n is a prime number.

Solution. We shall prove the given statement by contradiction.
Suppose that n is a composite number. Then $n = qm$, where q is a prime number and $m > 1$ is a positive integer.
Since $3^n - 2^n$ is a power of prime number, it follows that $3^n - 2^n = p^a$, where p is an odd prime number and a is a positive integer. We obtain

$$p^a = 3^{qm} - 2^{qm}$$
$$= (3^m)^q - (2^m)^q.$$

It follows that $3^m - 2^m$ divides p^a. Hence, $3^m - 2^m = p^b$, where b is a positive integer such that $1 < b < a$. Then $3^m = 2^m + p^b$. We obtain

$$p^a = (2^m + p^b)^q - 2^{qm}$$
$$= 2^{qm} + qp^b 2^{m(q-1)} + \dots + q2^m p^{b(q-1)} + p^{bq} - 2^{qm}$$
$$= qp^b 2^{m(q-1)} + \dots + q2^m p^{b(q-1)} + p^{bq}.$$

Since p^a is divisible by p^{b+1}, it implies that $p \mid q2^{m(q-1)}$. By knowing that p is an odd prime number, then $p \mid q$. Hence, $p = q$.
Consequently, $n = qm = pm$. As a result,

$$p^a = 3^{mp} - 2^{mp}$$

$$= (3^p)^m - (2^p)^m.$$

It follows that $3^p - 2^p$ divides p^a. Then $3^p - 2^p = p^c$ for some positive integers c. Hence,

$$3^p \equiv 2^p \pmod{p}.$$

Using Fermat's little theorem, we have $2^p \equiv 2 \pmod{p}$ and $3^p \equiv 3 \pmod{p}$. Thus, $3 \equiv 2 \pmod{p}$, a contradiction. Therefore, n is a prime number.

Problem 66. Evaluate the following expressions:

1. $\varphi(4)$ 5. $\varphi(16)$

2. $\varphi(6)$ 6. $\varphi(18)$

3. $\varphi(10)$ 7. $\varphi(20)$

4. $\varphi(12)$ 8. $\varphi(35)$

Solution. Evaluate the following expressions:

1. $\varphi(4) = \varphi(2^2) = 4\left(1 - \dfrac{1}{2}\right) = 2$

2. $\varphi(6) = \varphi(2 \times 3) = 6\left(1 - \dfrac{1}{2}\right)\left(1 - \dfrac{1}{3}\right) = 2$

3. $\varphi(10) = \varphi(2 \times 5) = 10\left(1 - \dfrac{1}{2}\right)\left(1 - \dfrac{1}{5}\right) = 4$

4. $\varphi(12) = \varphi(2^2 \times 3) = 12\left(1 - \dfrac{1}{2}\right)\left(1 - \dfrac{1}{3}\right) = 4$

5. $\varphi(16) = \varphi(2^4) = 16\left(1 - \dfrac{1}{2}\right) = 8$

6. $\varphi(18) = \varphi(2 \times 3^2) = 18\left(1 - \dfrac{1}{2}\right)\left(1 - \dfrac{1}{3}\right) = 6$

7. $\varphi(20) = \varphi(2^2 \times 5) = 20\left(1 - \dfrac{1}{2}\right)\left(1 - \dfrac{1}{5}\right) = 8$

8. $\varphi(35) = \varphi(5 \times 7) = 35\left(1 - \dfrac{1}{5}\right)\left(1 - \dfrac{1}{7}\right) = 24$

Problem 67. Find the last digits of the following numbers:

1. 3^{2025};   2. 7^{2026};   3. 13^{2027};   4. 17^{2028}.

Solution. Find the last digits of the following numbers:

1. 3^{2025}

   To determine the last digit of 3^{2025}, we need to find the remainder when 3^{2025} is divided by 10. Since $\gcd(10, 3) = 1$, using Euler's theorem, we obtain

   $$3^{\phi(10)} \equiv 1 \pmod{10}.$$

   It follows that
   $$3^4 \equiv 1 \pmod{10}.$$

   We have

   $$3^{2025} = \left(3^4\right)^{506} \times 3$$
   $$\equiv (1)^{506} \times 3 \pmod{10}$$
   $$\equiv 3 \pmod{10}.$$

   Therefore, 3 is the last digit of 3^{2025}.

2. 7^{2026}

   To determine the last digit of 7^{2026}, we need to find the remainder when 7^{2026} is divided by 10. Since $\gcd(10, 7) = 1$, using Euler's theorem, we obtain

   $$7^{\phi(10)} \equiv 1 \pmod{10}.$$

   It follows that
   $$7^4 \equiv 1 \pmod{10}.$$

   We have

   $$7^{2026} = \left(7^4\right)^{506} \times 7^2$$
   $$\equiv (1)^{506} \times 49 \pmod{10}$$
   $$\equiv 9 \pmod{10}.$$

   Therefore, 9 is the last digit of 7^{2026}.

3. 13^{2027}

   To determine the last digit of 13^{2027}, we need to find the remainder when 13^{2027} is divided by 10. Since $\gcd(10,13) = 1$, using Euler's theorem, we obtain

$$13^{\phi(10)} \equiv 1 \pmod{10}.$$

It follows that

$$13^4 \equiv 1 \pmod{10}.$$

We have

$$13^{2027} = \left(13^4\right)^{506} \times 13^3$$
$$\equiv (1)^{506} \times 3^3 \pmod{10}$$
$$\equiv 7 \pmod{10}.$$

Therefore, 7 is the last digit of 7^{2026}.

4. 17^{2028}

To determine the last digit of 17^{2028}, we need to find the remainder when 17^{2028} is divided by 10. Since $\gcd(10,17) = 1$, using Euler's theorem, we obtain

$$17^{\phi(10)} \equiv 1 \pmod{10}.$$

It follows that

$$17^4 \equiv 1 \pmod{10}.$$

We have

$$17^{2028} = \left(17^4\right)^{506} \times 17^4$$
$$\equiv (1)^{506} \times (-3)^4 \pmod{10}$$
$$\equiv 1 \pmod{10}.$$

Therefore, 1 is the last digit of 17^{2028}.

Problem 68. Find the remainder when

1. 4^{7777} is divided by 17;

2. 5^{8888} is divided by 17;

3. 6^{9999} is divided by 17;

4. 7^{6666} is divided by 17.

Solution. Find the remainder when

1. 4^{7777} is divided by 17
   Since $\gcd(4, 17) = 1$, using Euler's theorem, we have

   $$4^{\phi(17)} \equiv 1 \pmod{17}.$$

   It follows that
   $$4^{16} \equiv 1 \pmod{17}.$$

   We have

   $$\begin{aligned}
   4^{7777} &= \left(4^{16}\right)^{486} \times 4 \\
   &\equiv 1^{486} \times 4 \pmod{17} \\
   &\equiv 4 \pmod{17}.
   \end{aligned}$$

   Therefore, 4 is the remainder when 4^{7777} is divided by 17.

2. 5^{8888} is divided by 17
   Since $\gcd(5, 17) = 1$, using Euler's theorem, we obtain

   $$5^{\phi(17)} \equiv 1 \pmod{17}.$$

   Then
   $$5^{16} \equiv 1 \pmod{17}.$$

   We have

   $$\begin{aligned}
   5^{8888} &= \left(5^{16}\right)^{555} \times 5^8 \\
   &\equiv 1^{555} \times 25^4 \pmod{17} \\
   &\equiv 1 \times 8^4 \pmod{17} \\
   &\equiv 64^2 \pmod{17} \\
   &\equiv 13^2 \pmod{17} \\
   &\equiv 169 \pmod{17} \\
   &\equiv 16 \pmod{17}.
   \end{aligned}$$

   Therefore, 16 is the remainder when 5^{8888} is divided by 17.

3. 6^{9999} is divided by 17
   We have $\gcd(6, 17) = 1$. Using Euler's theorem, we obtain

   $$6^{\varphi(17)} \equiv 1 \pmod{17}.$$

It follows that
$$6^{16} \equiv 1. \quad (\text{mod } 17)$$

We have
$$6^{9999} = \left(6^{16}\right)^{624} \times 6^{15}$$
$$\equiv 1^{624} \times 6 \times 36^7 \quad (\text{mod } 17)$$
$$\equiv 6 \times 2^7 \quad (\text{mod } 17)$$
$$\equiv 3 \quad (\text{mod } 17).$$

Therefore, 3 is the remainder when 6^{9999} is divided by 17.

4. 7^{6666} is divided by 17.

Since $\gcd(17, 7) = 1$, using Euler's theorem, we obtain
$$7^{\varphi(17)} \equiv 1 \quad (\text{mod } 17).$$

Then
$$7^{16} \equiv 1 \quad (\text{mod } 17).$$

We have
$$7^{6666} = \left(7^{16}\right)^{416} \times 7^{10}$$
$$\equiv 1^{416} \times 49^5 \quad (\text{mod } 17)$$
$$\equiv (-2)^5 \quad (\text{mod } 17)$$
$$\equiv -32 \quad (\text{mod } 17)$$
$$\equiv 2 \quad (\text{mod } 17).$$

Therefore, 2 is the remainder when 7^{6666} is divided by 17.

Problem 69. Let $n_1, n_2, ..., n_k$ be k positive integers and a be an integer such that $\gcd(n_i, a) = 1$ for all $1 \leq n \leq k$. Prove that
$$a^{\text{lcm}(\phi(n_1), \phi(n_2), ..., \phi(n_k))} \equiv 1 \quad (\text{mod } \text{lcm}(n_1, n_2, ..., n_k)).$$

Solution. Since $\gcd(n_1, a) = 1$, using Euler's theorem, we obtain
$$a^{\phi(n_1)} \equiv 1 \quad (\text{mod } n_1).$$

By knowing that $\phi(n_1) \mid \text{lcm}(\phi(n_1), \phi(n_2), ..., \phi(n_k))$, we obtain
$$a^{\text{lcm}(\phi(n_1), \phi(n_2), ..., \phi(n_k))} \equiv 1 \quad (\text{mod } n_1). \tag{1}$$

Similarly, we obtain

$$a^{\text{lcm}(\phi(n_1),\phi(n_2),...,\phi(n_k))} \equiv 1 \pmod{n_2} \tag{2}$$

and

$$a^{\text{lcm}(\phi(n_1),\phi(n_2),...,\phi(n_k))} \equiv 1 \pmod{n_k}. \tag{k}$$

From (1), (2), ..., (k), it follows that

$$a^{\text{lcm}(\phi(n_1),\phi(n_2),...,\phi(n_k))} \equiv 1 \pmod{\text{lcm}(n_1, n_2, ..., n_k)}.$$

Problem 70. For all positive integers $n > 2$, prove that $\varphi(n)$ is an even number.

Solution. To prove the given statement, we study the following cases:

- If n is a power of 2, we obtain $n = 2^k$ for some $k \geq 2$. It follows that

$$\phi(n) = \phi\left(2^k\right)$$
$$= 2^k \left(1 - \frac{1}{2}\right)$$
$$= 2^k \times \frac{1}{2}$$
$$= 2^{k-1}.$$

Hence, $\varphi(n)$ is an even number.

- If n is not a power of 2, then $n = p^k m$, where p is an odd prime number and $\gcd(p, n) = 1$. Since p is multiplicative, it turns out that

$$\phi\left(p^k m\right) = \phi\left(p^k\right)\phi(m)$$
$$= p^k \left(1 - \frac{1}{p}\right)\phi(m)$$
$$= p^{k-1}(p-1)\phi(m).$$

Since p is an odd number, then $p - 1$ is an even number. Hence, $\varphi(n)$ is an even number.

Problem 71. Given n is a positive integer. Prove that the sum of the positive integers less than n and relatively prime to n is $\frac{1}{2}n\varphi(n)$.

Solution. Let $x_1, x_2, ..., x_{\varphi(n)}$ be the positive integers less than n and relatively prime to n. It is trivial to see that $\gcd(x, n) = \gcd(n - x, n)$. Hence, $\gcd(x, n) = 1$ if and only if $\gcd(n - x, n) = 1$. It turns out that $n - x_1, n - x_2, ..., n - x_{\varphi(n)}$ are equal to $x_1, x_2, ..., x_{\varphi(n)}$ in some order. Consequently,

$$x_1 + x_2 + ... + x_{\phi(n)} = (n - x_1) + (n - x_2) + ... + \left(n - x_{\varphi(n)}\right).$$

Then $2\left(x_1 + x_2 + ... + x_{\varphi(n)}\right) = n\phi(n)$.
Therefore,

$$x_1 + x_2 + ... + x_{\varphi(n)} = \frac{1}{2}n\varphi(n).$$

Problem 72. Let p be an odd prime number and a be an integer such that $\gcd(a, p) = 1$. Prove that

$$a^{\frac{p-1}{2}} \equiv \pm 1 \pmod{p}.$$

Solution. We have

$$a^p - 1 = \left(a^{\frac{p-1}{2}}\right)^2 - 1$$
$$= \left(a^{\frac{p-1}{2}} - 1\right)\left(a^{\frac{p-1}{2}} + 1\right).$$

Since p is an odd prime number and $\gcd(a, p) = 1$, using Euler's theorem, we obtain

$$a^{\varphi(p)} \equiv 1 \pmod{p}.$$

Then

$$a^{p-1} \equiv 1 \pmod{p}.$$

It implies that p divides $a^{p-1} - 1$.
Hence, p divides $\left(a^{\frac{p-1}{2}} - 1\right)\left(a^{\frac{p-1}{2}} + 1\right)$.
We obtain $p \mid a^{\frac{p-1}{2}} - 1$ or $p \mid a^{\frac{p-1}{2}} + 1$. It follows that

$$a^{\frac{p-1}{2}} \equiv 1 \pmod{p}$$

or

$$a^{\frac{p-1}{2}} \equiv -1 \pmod{p}.$$

Therefore, $a^{\frac{p-1}{2}} \equiv \pm 1 \pmod{p}$.

Problem 73. Find all positive integers x, y and z such that

$$\varphi(x-1) + \varphi(y-2) + \varphi(z-3) = 3.$$

Solution. For all positive integers x, we obtain $\varphi(x)$ is a positive integer. Hence, $\varphi(x-1) + \varphi(y-2) + \varphi(z-3) = 3$ if and only if

$$\phi(x-1) = \phi(y-2) = \phi(z-3) = 1.$$

Then $\begin{cases} x - 1 = 1 \text{ or } 2 \\ y - 2 = 1 \text{ or } 2 \\ z - 3 = 1 \text{ or } 2 \end{cases}$. It follows that $\begin{cases} x = 2 \text{ or } 3 \\ y = 3 \text{ or } 4 \\ z = 4 \text{ or } 5 \end{cases}$.

Therefore, $(x, y, z) = (2, 3, 4), (2, 3, 5), (2, 4, 4), (2, 4, 5), (3, 3, 4), (3, 3, 5), (3, 4, 4), (3, 4, 5)$.

Problem 74. Given a and b are two positive integers such that $\gcd(a, b) = 1$. Prove that $a^{\phi(b)} + b^{\phi(a)} \equiv 1 \pmod{ab}$.

Solution. Since a and b are relatively prime, using Euler's theorem, we obtain

$$a^{\phi(b)} \equiv 1 \pmod{b}.$$

Moreover,

$$b^{\phi(a)} \equiv 0 \pmod{b}.$$

It follows that

$$a^{\phi(b)} + b^{\phi(a)} \equiv 1 \pmod{b}. \tag{1}$$

Similarly, using Euler's theorem, we have

$$b^{\phi(a)} \equiv 1 \pmod{a}$$

and

$$a^{\phi(b)} \equiv 0 \pmod{a}.$$

Then

$$a^{\phi(b)} + b^{\phi(a)} \equiv 1 \pmod{a}. \tag{2}$$

From (1) and (2), since $\gcd(a, b) = 1$, we obtain

$$a^{\phi(b)} + b^{\phi(a)} \equiv 1 \pmod{ab}.$$

Problem 75. Given $a \geq 2$ is an integer and n is a positive integer. Prove that $\varphi(a^n - 1)$ is divisible by n.

Solution. Since a and $a^n - 1$ are relatively prime, using Euler's theorem, we obtain

$$a^{\phi(a^n-1)} \equiv 1 \pmod{a^n - 1}.$$

It follows that $a^n - 1$ divides $a^{\phi(a^n-1)} - 1$. Therefore, $\varphi(a^n - 1)$ is divisible by n.

Problem 76. Given p and q are two distinct prime numbers. Let a be an integer such that $\gcd(a, p) = \gcd(a, q) = 1$. Prove that

$$p^2 a^{q(q-1)} + q^2 a^{p(p-1)} \equiv p^2 + q^2 \pmod{p^2 q^2}.$$

Solution. Since $\gcd(a, p) = 1$, then $\gcd(a, p^2) = 1$. Using Euler's theorem, we obtain

$$a^{\phi(p^2)} \equiv 1 \pmod{p^2}.$$

It follows that

$$a^{p(p-1)} \equiv 1 \pmod{p^2}.$$

Then

$$q^2 a^{p(p-1)} \equiv q^2 \pmod{p^2 q^2}. \tag{1}$$

Similarly, we obtain

$$p^2 a^{q(q-1)} \equiv p^2 \pmod{p^2 q^2}. \tag{2}$$

Adding (1) and (2), we obtain

$$p^2 a^{q(q-1)} + q^2 a^{p(p-1)} \equiv p^2 + q^2 \pmod{p^2 q^2}.$$

Problem 77. Find the remainder when 85! is divided by 89.

Solution. Since 89 is a prime number, using Wilson's theorem, we obtain

$$(89 - 1)! \equiv -1 \pmod{89}.$$

It follows that

$$88! \equiv -1 \pmod{89}$$
$$88 \times 87 \times 86 \times 85! \equiv -1 \pmod{89}$$
$$(-1)(-2)(-3) \times 85! \equiv -1 \pmod{89}$$
$$-6 \times 85! \equiv -1 \pmod{89}$$

$$6 \times 85! \equiv 1 \pmod{89}$$
$$90 \times 85! \equiv 15 \pmod{89}$$
$$85! \equiv 15 \pmod{89}$$

Therefore, 15 is the remainder when 85! is divided by 89.

Problem 78. Find the remainder when 104! is divided by 109.

Solution. Since 109 is a prime number, by Wilson's theorem, we obtain
$$(109 - 1)! \equiv -1 \pmod{109}.$$
It follows that
$$108! \equiv -1 \pmod{109}$$
$$108 \times 107 \times 106 \times 105 \times 104! \equiv -1 \pmod{109}$$
$$(-1)(-2)(-3)(-4) \times 104! \equiv -1 \pmod{109}$$
$$24 \times 104! \equiv -1 \pmod{109}$$
$$50 \times 24 \times 104! \equiv -50 \pmod{109}$$
$$104! \equiv 59 \pmod{109}$$

Therefore, 59 is the remainder when 104! is divided by 109.

Problem 79. Prove that $(41!)^2 - 1517 \times 40! + 38 \equiv 0 \pmod{2021}$.

Solution. Observe that $(41!)^2 - 1517 \times 40! + 38 \equiv 0 \pmod{2021}$ is equivalent to
$$(41!)^2 - 37 \times 41! + 38 \equiv 0 \pmod{2021}.$$

It is equivalent to $(41! - 1)(41! - 38) \equiv 0 \pmod{43 \times 47}$. Hence, to prove the given statement, we shall show that $43 \mid 41! - 1$ and $47 \mid 41! - 38$.

- Since 43 is a prime number, using Wilson's theorem, we obtain $(43 - 1)! \equiv -1 \pmod{43}$. Then $42! \equiv -1 \pmod{43}$. It follows that
$$42 \times 41! \equiv -1 \pmod{43}$$
$$(-1) \times 41! \equiv -1 \pmod{43}$$
$$41! \equiv 1 \pmod{43}.$$

Consequently,
$$43 \mid 41! - 1. \tag{1}$$

- Since 47 is a prime number, using Wilson's theorem, we obtain $(47 - 1)! \equiv -1 \pmod{47}$. It follows that

$$46! \equiv -1 \pmod{47}$$
$$46 \times 45 \times 44 \times 43 \times 42 \times 41! \equiv -1 \pmod{47}$$
$$(-1)(-2)(-3)(-4)(-5) \times 41! \equiv -1 \pmod{47}$$
$$-120 \times 41! \equiv -1 \pmod{47}$$
$$120 \times 41! \equiv 1 \pmod{47}$$
$$38 \times 120 \times 41! \equiv 38 \pmod{47}$$
$$41! \equiv 38 \pmod{47}$$

It turns out that
$$47 \mid 41! - 38. \tag{2}$$
Since $\gcd(43, 47) = 1$, from (1) and (2), we obtain
$$2021 \mid (41! - 1)(41! - 38).$$

Therefore, the given statement is proved.

Problem 80. Suppose that p is a prime number. Prove that

$$(p - 2)! \equiv 1 \pmod{p}.$$

Solution. Since p is a prime number, using Wilson's theorem, we obtain
$$(p - 1)! \equiv -1 \pmod{p}.$$
It follows that $(p - 1)! \equiv p - 1 \pmod{p}$.
By knowing that $\gcd(p - 1, p) = 1$, divide both sides of the modulo congruence by $p - 1$, we obtain $(p - 2)! \equiv 1 \pmod{p}$.

Problem 81. Let p be an odd prime. Prove the following statements:

$$1^2 \times 3^2 \times 5^2 \times ... \times (p - 2)^2 \equiv (-1)^{\frac{p+1}{2}} \pmod{p}.$$

Solution. Since p is a prime number, using Wilson's theorem, we obtain $(p - 1)! \equiv -1 \pmod{p}$. Moreover, $k \equiv -(p - k) \pmod{p}$. It follows that

$$1 \times 3 \times 5 \times ... \times (p - 2) \equiv (-1)^{\frac{p-1}{2}} 2 \times 4 \times 6 \times ... \times (p - 1) \pmod{p}.$$

Multilpy both sides of the congruence by $1 \times 3 \times 5 \times ... \times (p-2)$, we obtain

$$1^2 \times 3^2 \times 5^2 \times ... \times (p-2)^2 \equiv (-1)^{\frac{p-1}{2}} (p-1)! \pmod{p}.$$

It turns out that

$$
\begin{aligned}
1^2 \times 3^2 \times 5^2 \times ... \times (p-2)^2 &\equiv (-1)^{\frac{p-1}{2}} (-1) \\
&\equiv (-1)^{\frac{p-1}{2}+1} \\
&\equiv (-1)^{\frac{p+1}{2}} \pmod{p}.
\end{aligned}
$$

Therefore, $1^2 \times 3^2 \times 5^2 \times ... \times (p-2)^2 \equiv (-1)^{\frac{p+1}{2}} \pmod{p}$.

Problem 82. Is $171 \times 534! - 1$ a composite number?

Solution. It is easy to check that 541 is a prime number. Using Wilson's theorem, we obtain

$$
\begin{aligned}
540! &\equiv -1 \\
540 \times 539 \times 538 \times 537 \times 536 \times 535 \times 534! &\equiv -1 \\
(-1)(-2)(-3)(-4)(-5)(-6)(-7) \times 534! &\equiv -1 \\
-7! \times 534! &\equiv -1 \\
5040 \times 534! &\equiv 1 \\
171 \times 534! - 1 &\equiv 0
\end{aligned}
$$

Hence, $541 \mid 171 \times 534! - 1$.
Therefore, $171 \times 534! - 1$ is a composite number.

Problem 83. (Lerch's Congruence)
Given p is an odd prime. Prove that

$$1^{p-1} + 2^{p-1} + ... + (p-1)^{p-1} \equiv p + (p-1)! \pmod{p^2}.$$

Solution. Using Fermat's little theorem, we obtain $i^{p-1} \equiv 1 \pmod{p}$. Hence, $i^{p-1} = 1 + pk_i$ for some integers k_i. We obtain

$$
\begin{aligned}
&1^{p-1} + 2^{p-1} + ... + (p-1)^{p-1} \\
&= (1 + pk_1) + (1 + pk_2) + ... + (1 + pk_{p-1}) \\
&= p - 1 + p(k_1 + k_2 + ... + k_{p-1}).
\end{aligned} \tag{i}
$$

Moreover,

$$(p-1)!^{p-1} = (1+pk_1)(1+pk_2)\dots(1+pk_{p-1})$$
$$\equiv 1 + p(k_1 + k_2 + \dots + k_{p-1}) \pmod{p^2}. \quad (1)$$

Using Wilson's theorem, we have $(p-1)! \equiv -1 \pmod{p}$.
Then $(p-1)! = -1 + pk$ for some integer k. It follows that

$$(p-1)!^{p-1} = (-1+pk)^{p-1}$$
$$\equiv (-1)^{p-1} + (p-1)(-1)^{p-2}pk \pmod{p^2}$$
$$\equiv 1 - (p-1)pk \pmod{p^2}$$
$$\equiv 1 - p^2 k + pk \pmod{p^2}$$
$$\equiv 1 + pk \pmod{p^2}. \quad (2)$$

From (1) and (2), we obtain

$$1 + p(k_1 + k_2 + \dots + k_{p-1}) \equiv 1 + pk.$$

Then

$$p(k_1 + k_2 + \dots + k_{p-1}) \equiv pk.$$

From (i), it implies that

$$1^{p-1} + 2^{p-1} + \dots + (p-1)^{p-1} \equiv p - 1 + pk \pmod{p^2}$$
$$\equiv p + (p-1)! \pmod{p^2}.$$

Therefore, $1^{p-1} + 2^{p-1} + \dots + (p-1)^{p-1} \equiv p + (p-1)! \pmod{p^2}$.

Problem 84. Given a prime number p and an integer n such that $0 \le n \le p-1$. Prove that

$$n!\,(p-n-1)! + (-1)^n \equiv 0 \pmod{p}.$$

Solution. Observe that

$$p - k - 1 \equiv -(k+1) \pmod{p}$$

for all integers k such that $0 \le k \le p-1$.
Hence,

$$(p-n-1)! \equiv (-1)^{p-n-1}(n+1)(n+2)\dots(p-1) \pmod{p}.$$

It follows that

$$n!\,(p-n-1)! \equiv (-1)^{p-1}(-1)^{-n}\,(p-1)! \pmod{p}.$$

Since p is a prime number, using Wilson's theorem, we obtain

$$(p-1)! \equiv -1 \pmod{p}.$$

It turns out that

$$\begin{aligned}
n!\,(p-n-1)! &\equiv (-1)^{p-1}(-1)^n\,(-1) \pmod{p} \\
&\equiv -(-1)^n \pmod{p}.
\end{aligned}$$

Therefore, $n!\,(p-n-1)! + (-1)^n \equiv 0 \pmod{p}$.

Problem 85. Given x is an integer such that

$$1 + \frac{1}{2} + \frac{1}{3} + \dots + \frac{1}{2017} = \frac{x}{2017!}.$$

Prove that $x \equiv 1 \pmod{1009}$.

Solution. We have $1 + \dfrac{1}{2} + \dfrac{1}{3} + \dots + \dfrac{1}{2017} = \dfrac{x}{2017!}$.
Multiply both sides of the equality by $2017!$, we obtain

$$2017! + \frac{2017!}{2} + \frac{2017!}{3} + \dots + \frac{2017!}{2017} = x.$$

It turns out that $x \equiv \dfrac{2017!}{1009} \pmod{1009}$.
Then

$$\begin{aligned}
x &\equiv 1 \times 2 \times \dots \times 1008 \times 1010 \times \dots \times 2017 \\
&\equiv 1008! \times 1 \times 2 \times \dots \times 1008 \\
&\equiv 1008! \times 1008! \\
&\equiv (1008!)^2 \pmod{1009}.
\end{aligned}$$

Since 1009 is a prime number, using Wilson's theorem, we have $1008! \equiv -1 \pmod{1009}$. It implies that $x \equiv (-1)^2 \pmod{1009}$. Therefore, $x \equiv 1 \pmod{1009}$.

Problem 86. Given a prime number p. Prove that

$$(p-1)! \equiv p-1 \pmod{p(p-1)}.$$

69

Solution. Using Wilson's theorem, we have

$$(p-1)! \equiv -1. \pmod{p}.$$

It follows that

$$(p-1)! \equiv p-1 \pmod{p}.$$

Moreover,

$$(p-1)! \equiv p-1 \pmod{p-1}.$$

Since $\gcd(p, p-1) = 1$, we obtain

$$(p-1)! \equiv p-1 \pmod{p(p-1)}.$$

Problem 87. Given two positive integers a and b such that $a \geq 2$ and $\gcd(a, b) = 1$. Prove that

$$b^{a-1} + (a-1)! \equiv 0 \pmod{a}$$

if and only if a is a prime number.

Solution. $\Rightarrow$ Suppose that $b^{a-1} + (a-1)! \equiv 0 \pmod{a}$. We shall show that a is a prime number. We will prove this statement by contradiction. Suppose that a is composite. That is, $a = mn$, where m and n are two positive integers greater than 1.
Since $a \mid b^{a-1} + (a-1)!$. We obtain $m \mid b^{a-1} + (a-1)!$.
Since $m \mid (a-1)!$, it implies that $m \mid b^{a-1}$.
We know that $\gcd(a, b) = 1$. Then $\gcd(m, a) = 1$.
Hence, $m = 1$, a contradiction.
Therefore, a is a prime number.
$\Leftarrow$ Suppose that a is a prime number.
From Fermat's little theorem and Wilson's theorem, we obtain

$$b^{a-1} \equiv 1 \pmod{a}$$

and

$$(b-1)! \equiv -1 \pmod{a}.$$

Consequently, $b^{a-1} + (a-1)! \equiv 0 \pmod{a}$.

Problem 88. Find the greatest common divisor of the following integers:

1. 18 and 36   3. 45 and 57   5. 200 and 150

2. 20 and 30   4. 108 and 99   6. 198 and 196

Solution. Find the greatest common divisor of the following integers:

1. 18 and 36

   - $18 = 2 \times 3^2$
   - $36 = 2^2 \times 3^2$

   It turns out that $\gcd(18, 36) = 2 \times 3^2 = 18$.

2. 20 and 30

   - $20 = 2^2 \times 5$
   - $30 = 2 \times 3 \times 5$

   It turns out that $\gcd(20, 30) = 2 \times 5 = 10$.

3. 45 and 57

   - $45 = 3^2 \times 5$
   - $57 = 3 \times 19$

   It turns out that $\gcd(45, 57) = 3$.

4. 108 and 99

   - $108 = 3^2 \times 13$
   - $99 = 3^2 \times 11$

   Consequently, $\gcd(108, 99) = 9$.

5. 200 and 150

   - $200 = 2^3 \times 5^2$
   - $150 = 2 \times 3 \times 5^2$

   Hence, $\gcd(200, 150) = 2 \times 5^2 = 50$.

6. 198 and 196

   - $198 = 2 \times 3^2 \times 11$

- $196 = 2^2 \times 7^2$

Consequently, $\gcd(198, 196) = 2$.

Problem 89. Prove that there are no integers x and y such that $x + y = 11$ and $\gcd(x, y) = 4$.

Solution. Since $\gcd(x, y) = 4$, it follows that $x = 4m$ and $y = 4n$ for some integers m and n such that $\gcd(m, n) = 1$. Hence,

$$x + y = 11$$

implies that

$$4m + 4n = 11.$$

Then $4(m + n) = 11$.
It turns out that $4 \mid 11$, a contradiction.
Therefore, there are no integers x and y such that $x + y = 11$ and $\gcd(x, y) = 4$.

Problem 90. Let a, b and k be two positive integers. Suppose that $\gcd(\gcd(a, a + kb), k) = 1$. Prove that

$$\gcd(a, a + kb) = \gcd(a, b).$$

Solution. Prove that $\gcd(a, a + kb) = \gcd(a, b)$.
Let $d = \gcd(a, b)$ and $d' = \gcd(a, a + kb)$. We shall show that $d = d'$.
Since $d = \gcd(a, b)$, it follows that $d \mid a$ and $d \mid b$.
It turns out that $d \mid a + kb$.
Then d is a common divisor of a and $a + kb$.
By definition, we obtain

$$d \leq d'. \tag{1}$$

Moreover, $d' = \gcd(a, a + kb)$. Then $d' \mid a$ and $d' \mid a + kb$.
Then $d' \mid (a + kb) - a = kb$.
Since $\gcd(\gcd(a, a + kb), k) = 1$, we obtain $d' \mid b$.
It turns out that d' is a common divisor of a and b.
Consequently,

$$d' \leq d. \tag{2}$$

From (1) and (2), we obtain $d = d'$.
Therefore, $\gcd(a, a + kb) = \gcd(a, b)$.

Problem 91. A linear combination of two positive integers a and b is the sum of the multiples of a and b. That is, the linear combination of two positive integers a and b is an integer of the form $ax + by$, where x and y are integers. Prove that $\gcd(a, b)$ is a linear combination of a and b.

Solution. We consider on the set of positive linear combination of a and b which is defined by

$$L = \{am + bn \mid m, n \in \mathbb{Z}, am + bn > 0\}.$$

We know that $a > 0$ and $a = 1 \times a + 0 \times b$. Then $a \in L$. It turns out that L is not an empty set. That is, L is a nonempty subset of $\mathbb{N}$. By Well-ordering property, L has the least element, say d. We shall show that $d = \gcd(a, b)$. We first prove that d is a common divisor of a and b. Using division algorithm, we have $a = dq + r$, where $0 \leq r < d$. It follows that $r = a - dq$. Since $d \in L$, then there exists x and $y \in \mathbb{Z}$ such that $d = ax + by$. It follows that

$$r = a - q(ax + by)$$
$$= a - aqx - bqy$$
$$= a(1 - qx) - qyb.$$

If $r > 0$, we obtain r is a positive linear combination of a and b. Then $r \in L$. Hence, d is not the least element of L, a contradiction. It turns out that $r = 0$. Consequently, $a = dq$. That is, $d \mid a$. Similarly, we obtain $b \mid d$. It follows that d is a common divisor of a and b. Suppose that d' is a positive common divisor a and b. Then $d' \mid a$ and $d' \mid b$. We obtain $d' \mid ax + by = d$. It implies that $d' \leq d$. Thus, $d = \gcd(a, b)$. Therefore, $\gcd(a, b)$ is a linear combination of a and b.

Problem 92. Suppose that a and b are two positive integers. Let $d = \gcd(a, b)$ and d' is a common divisor of a and b. Prove that $d' \mid d$.

Solution. Since $d = \gcd(a, b)$, then d is the least positive linear combination of a and b. That is, there exist integers x and y such that $d = ax + by$. Moreover, d' is a common divisor of a and b. It follows that $d' \mid a$ and $d' \mid b$. It turns out that $d' \mid ax + by$. Therefore, $d' \mid d$.

73

Problem 93. Let a and b be two positive integers. Suppose that $a = dm$ and $b = dn$. Prove that $d = \gcd(a,b)$ if and only if $\gcd(m,n) = 1$.

Solution. $\Rightarrow$ Suppose that $d = \gcd(a,b)$. We shall show that $\gcd(m,n) = 1$. Assume that $d' = \gcd(m,n)$. Then $m = d'k$ and $n = d'l$ for some integers k and l. It follows that $a = dd'k$ and $b = dd'l$. It turns out that dd' is a common divisor of a and b. By definition, we obtain $dd' \leq d$. Then $d' \leq 1$. Consequently, $d' = 1$. Therefore, $\gcd(m,n) = 1$.
$\Leftarrow$ Suppose that $\gcd(m,n) = 1$. We will show that $d = \gcd(a,b)$. Since $a = dm$ and $b = dn$, it follows that $\gcd(a,b) = dl$. It turns out that $dl \mid dm$. Then $l \mid m$.
Similarly, $l \mid n$. Hence, l is a common divisor of m and n. As a result, $l \leq 1$ since $\gcd(m,n) = 1$. It implies that $l = 1$. Thus, $\gcd(a,b) = d$.

Problem 94. Given a, b and k are three positive integers. Prove that

$$\gcd(ak, bk) = k \gcd(a,b).$$

Solution. Let $d = \gcd(a,b)$. Then $a = dm$ and $b = dn$ for some positive integers m and n such that $\gcd(m,n) = 1$. We obtain $ka = kdm$ and $kb = kdn$, where $\gcd(m,n) = 1$.
Therefore, $\gcd(ka, kb) = kd = k \gcd(a,b)$.

Problem 95. Given two positive integers a and b. Prove that a and b are relatively prime if and only if $ax + by = 1$ for some integers x and y.

Solution. $\Rightarrow$ Suppose that a and b are relatively prime. We shall show that $ax + by = 1$. Since a and b are relatively prime, then $\gcd(a,b) = 1$. Then there exist integers x and y such that

$$ax + by = 1.$$

$\Leftarrow$ Suppose that $ax + by = 1$ for some integers x and y. We shall show that a and b are relatively prime.
Let $d = \gcd(a,b)$. Then $d \mid a$ and $d \mid b$. It follows that $d \mid ax + by = 1$. Hence, $d = 1$. Therefore, a and b are relatively prime.

Problem 96. Prove that every two consecutive positive integers are relatively prime.

Solution. Let n and $n + 1$ be the given two consecutive positive integers. We see that $(-1)\,n + (1)\,(n + 1) = 1$. Therefore, n and $n + 1$ are relatively prime.

Problem 97. Prove that the following fractions are irreducible for all positive integers n:

1. $\dfrac{n + 1}{7n + 6}$

2. $\dfrac{n\,(n + 2)}{n + 1}$

3. $\dfrac{n + 1}{n^2 + 4n + 2}$

4. $\dfrac{n + 1}{n\,(n^2 + 3n + 3)}$

Solution. Prove that the following fractions are irreducible for all positive integers n:

1. $\dfrac{n + 1}{7n + 6}$
We have
$$7\,(n + 1) - (7n + 6) = 7n + 7 - 7n - 6$$
$$= 1.$$

Then $n + 1$ and $7n + 6$ are relatively prime.
Therefore, $\dfrac{n + 1}{7n + 6}$ is irreducible.

2. $\dfrac{n\,(n + 2)}{n + 1}$
We have
$$(n + 1)\,(n + 1) - n\,(n + 2) = n^2 + 2n + 1 - n^2 - 2n$$
$$= 1.$$

Hence, $n + 1$ and $n(n + 2)$ are relatively prime.
Therefore, $\dfrac{n\,(n + 2)}{n + 1}$ is irreducible.

3. $\dfrac{n + 1}{n^2 + 4n + 2}$
We have
$$(n + 3)\,(n + 1) - \left(n^2 + 4n + 2\right)$$
$$= n^2 + 4n + 3 - n^2 - 4n - 2$$

75

$$= 1.$$

It follows that $n + 1$ and $n^2 + 4n + 2$ are relatively prime.
Therefore, $\dfrac{n + 1}{n^2 + 4n + 2}$ is irreducible.

4. $\dfrac{n + 1}{n\,(n^2 + 3n + 3)}$
We have

$$(n + 1)^2\,(n + 1) - n\,(n^2 + 3n + 3)$$
$$= n^3 + 3n^2 + 3n + 1 - n^3 - 3n^2 - 3n$$
$$= 1.$$

It turns out that $n+1$ and $n(n^2+3n+3)$ are relatively prime.
Therefore, $\dfrac{n + 1}{n\,(n^2 + 3n + 3)}$ is irreducible.

Problem 98. Given a, b and c are three positive integers such that $\gcd(a, b) = 1$ and $\gcd(a, c) = 1$. Prove that $\gcd(a, bc) = 1$.

Solution. Since $\gcd(a, b) = 1$ and $\gcd(a, c) = 1$, then there exist integers u, v, x, y such that $au + bv = 1$ and $ax + cy = 1$. Multiply the last two equations, we obtain

$$(au + bv)\,(ax + cy) = 1$$
$$a^2 ux + acuy + abvx + bcvy = 1$$
$$a\,(aux + cuy + bvx) + bc\,(vy) = 1$$
$$as + bct = 1$$

, where $s = aux + cuy + bvx$ and $t = vy$ are integers.
Therefore, $\gcd(a, bc) = 1$.

Problem 99. Suppose that a, b and c are three positive integers such that $\gcd\,(b, c) = 1$. Prove that $\gcd\,(a, b) = \gcd\,(ac, b)$.

Solution. Let $d = \gcd\,(a, b)$ and $d' = \gcd\,(ac, b)$.
Then $d' \mid ac$ and $d' \mid b$.
Using the fact that $\gcd\,(b, c) = 1$, then there exist an integer x and y such that $bx + cy = 1$.
Multiply both sides of the last equality by a, we obtain

$$abx + acy = a.$$

Hence, $d' \mid abx + acy = a$.
It turns out that d' is a common divisor of a and b.
Thus,

$$d' \mid d. \tag{1}$$

Moreover, $d \mid a$ and $d \mid b$. Then $d \mid ac$ and $d \mid b$.
It implies that d is a common divisor of ac and b.
Hence,

$$d \mid d'. \tag{2}$$

From (1) and (2), we obtain $d = d'$.
Therefore, $\gcd(a, b) = \gcd(ac, b)$.

Remark 3. Given four positive integers a, b, c and d such that $\gcd(a, c) = 1$ and $\gcd(b, d) = 1$. Then $\gcd(a, b) = \gcd(ad, bc)$.

Problem 100. Given two positive integers a and b such that $\gcd(a, b) = 1$. Prove that $\gcd(a^n, b^n) = 1$ for all positive integers n.

Solution. We will prove the given statement by using induction. For the case $n = 1$, we obtain $\gcd(a^n, b^n) = \gcd(a, b) = 1$, which is true. Suppose that $\gcd(a^k, b^k) = 1$ for some positive integers k. We shall show that $\gcd(a^{k+1}, b^{k+1}) = 1$.
From problem 99 and induction hypothesis, we have

$$\gcd(a^{k+1}, b^{k+1}) = \gcd(a \times a^k, b \times b^k)$$
$$= \gcd(a^k, b^k)$$
$$= 1.$$

Therefore, $\gcd(a^n, b^n) = 1$ for all positive integers n.

Problem 101. Given two positive integers a and b such that

$$\gcd(a, b) = 1.$$

Prove that $\gcd(a + b, ab) = 1$.

Solution. Let $d = \gcd(a + b, ab)$. Then $d \mid a + b$ and $d \mid ab$.
It follows that $d \mid b(a + b) - ab = b^2$.
Similarly, we obtain $d \mid a^2$.
Hence, d is a common divisor of a^2 and b^2.

It turns out that $d \mid \gcd(a^2, b^2)$.
From problem 100, we have

$$\gcd\left(a^2, b^2\right) = \left[\gcd\left(a, b\right)\right]^2 = 1.$$

Consequently, $d \mid 1$. We obtain $d = 1$.
Therefore, $\gcd(a + b, ab) = 1$.

Problem 102. Given two positive integers a and b such that

$$\gcd(a, b) = 1.$$

Prove that

    1. $\gcd\left(a + b, b^2\right) = 1$;   2. $\gcd\left(a + b, 3a^2 + b^2\right) = 1$.

Solution. Prove that

    1. $\gcd\left(a + b, b^2\right) = 1$
        Let $d = \gcd(a + b, b^2)$. Then $d \mid a + b$ and $d \mid b^2$.
        It follows that $d \mid b(a + b) - b^2 = ab$.
        It turns out that d is a common divisor of $a + b$ and ab.
        Hence, $d \mid \gcd(a + b, ab) = 1$.
        We obtain $d = 1$.
        Therefore, $\gcd\left(a + b, b^2\right) = 1$.

    2. $\gcd\left(a + b, 3a^2 + b^2\right) = 1$
        Since $\gcd(a, b) = 1$, then a and b have opposite parity. That
        is, $a + b$ is an odd number. Let $d = \gcd(a + b, 3a^2 + b^2)$.
        Then d is an odd number.
        Moreover, $d \mid a + b$ and $d \mid 3a^2 + b^2$.
        It follows that

$$\begin{aligned}
d \mid (3a + b)\,(a + b) &- \left(3a^2 + b^2\right) \\
&= 3a^2 + 4ab + b^2 - 3a^2 - b^2 \\
&= 4ab.
\end{aligned}$$

Then $d \mid ab$ since d is an odd number.
Hence, d is a common divisor of $a + b$ and ab.
It turns out that $d \mid \gcd(a + b, ab) = 1$. Then $d = 1$.
Therefore, $\gcd\left(a + b, 3a^2 + b^2\right) = 1$.

Problem 103. Given a, b, k and l are four positive integers such that $\gcd(a, b) = 1$ and $k + l = 2^m$ for some nonnegative integers m. Prove that $\gcd(a + b, ka^2 + lb^2) = 1$.

Solution. Since $\gcd(a, b) = 1$, then a and b have opposite parity. Let $d = \gcd(a + b, ka^2 + lb^2)$. It follows that d is an odd number. By definition, we have $d \mid a + b$ and $d \mid ka^2 + lb^2$. We obtain d is an odd number.

Moreover,

$$d \mid (ka + lb)(a + b) - (ka^2 + lb^2)$$
$$= ka^2 + (k + l) ab + lb^2 - ka^2 - lb^2$$
$$= (k + l) ab.$$

Since $k + l = 2^m$ and d is an odd number, we obtain $\gcd(d, k + l) = 1$. It turns out that $d \mid ab$. Hence, d is a common divisor of $a + b$ and ab. Consequently, $d \mid \gcd(a, b) = 1$. We obtain $d = 1$. Therefore, $\gcd(a + b, ka^2 + lb^2) = 1$.

Problem 104. Given two positive integers a and b such that

$$\gcd(a, b) = 1.$$

Prove that

1. $\gcd\left(a^2 + 3b^2, (a + b)^3 + 8b^3 + a + b\right) = 1$;

2. $\gcd\left(a^2 + 3b^2, a^4 + 4a^2b^2 + 3b^4 + a + b\right) = 1$.

Solution. Prove that

1. $\gcd\left(a^2 + 3b^2, (a + b)^3 + 8b^3 + a + b\right) = 1$;

   Let $d = \gcd\left(a^2 + 3b^2, (a + b)^3 + 8b^3 + a + b\right)$.

   Then $d \mid a^2 + 3b^2$ and $d \mid (a + b)^3 + 8b^3 + a + b$.

   It follows that

   $$d \mid (a + b)^3 + 8b^3 + a + b - (a + 3b)(a^2 + 3b^2)$$
   $$= (a + b)^3 + 8b^3 + a + b - a^3 - 3ab^2 - 3a^2b - 9b^3$$
   $$= (a + b)^3 + a + b - a^3 - 3ab^2 - 3a^2b - b^3$$
   $$= (a + b)^3 + a + b - (a + b)^3$$
   $$= a + b.$$

79

It turns out that d is a common divisor of $a+b$ and a^2+3b^2. Hence, $d \mid \gcd(a+b, a^2+3b^2)$. From problem 103, we have $\gcd(a+b, a^2+3b^2) = 1$. It implies that $d \mid 1$. Then $d = 1$. Therefore, $\gcd\left(a^2+3b^2, (a+b)^3 + 8b^3 + a + b\right) = 1$.

2. $\gcd\left(a^2+3b^2, a^4+4a^2b^2+3b^4+a+b\right) = 1$.
   Let $d = \gcd\left(a^2+3b^2, a^4+4a^2b^2+3b^4+a+b\right)$.
   Then $d \mid a^2+3b^2$ and $d \mid a^4+4a^2b^2+3b^4+a+b$.
   It follows that

   $$d \mid \left(a^4+4a^2b^2+3b^4+a+b\right) - \left(a^2+b^2\right)\left(a^2+3b^2\right)$$
   $$= \left(a^4+4a^2b^2+3b^4+a+b\right) - \left(a^4+4a^2b^2+3b^4\right)$$
   $$= a+b.$$

   Hence, d is a common divisor of $a+b$ and a^2+3b^2.
   It implies that $d \mid \gcd(a+b, a^2+3b^2)$.
   From problem ??, we have $\gcd(a+b, a^2+3b^2) = 1$.
   Then $d \mid 1$. It follows that $d = 1$.
   Therefore, $\gcd\left(a^2+3b^2, a^4+4a^2b^2+3b^4+a+b\right) = 1$.

Problem 105. Given two positive integers a and b such that

$$\gcd(a, b) = 1.$$

Prove that

1. $\gcd\left(a+b, a-b\right) = 1$ or 2;   2. $\gcd(a+b, 3a+2b) = 1$.

Solution. Prove that

1. $\gcd\left(a+b, a-b\right) = 1$ or 2;
   Let $d = \gcd(a+b, a-b)$. Then $d \mid a+b$ and $d \mid a-b$.
   It follows that

   $$d \mid (a+b) + (a-b) = 2a$$

   and

   $$d \mid (a+b) - (a-b) = 2b.$$

   Hence, d is a common divisor of $2a$ and $2b$.
   We obtain $d \mid \gcd(2a, 2b) = 2\gcd(a, b) = 2$.
   Then $d = 1$ or 2.
   Therefore, $\gcd\left(a+b, a-b\right) = 1$ or 2.

2. $\gcd(a + b, 3a + 2b) = 1$.
   Let $d = \gcd(a + b, 3a + 2b)$. Then $d \mid a + b$ and $d \mid 3a + 2b$.
   It follows that $d \mid 3(a + b) - (3a + 2b) = b$.
   Moreover, $d \mid (3a + 2b) - 2(a + b) = a$.
   It turns out that d is a common divisor of a and b.
   Hence, $d \mid \gcd(a, b) = 1$.
   Therefore, $\gcd(a + b, 3a + 2b) = 1$.

Problem 106. Given four positive integers a, b, x and y such that $\gcd(a, b) = \gcd(x, y)$. Prove that

$$\gcd\left(a^2, b^2\right) = \gcd\left(x^2, y^2\right).$$

Solution. From problem 100, if $\gcd(m, n) = 1$, we obtain

$$\gcd(m^2, n^2) = 1.$$

Let $d = \gcd(a, b)$. Then $\gcd\left(\dfrac{a}{d}, \dfrac{b}{d}\right) = 1$. It follows that

$$\gcd\left(\frac{a^2}{d^2}, \frac{b^2}{d^2}\right) = 1.$$

Hence,

$$
\begin{aligned}
\gcd\left(a^2, b^2\right) &= \gcd\left(d^2 \times \frac{a^2}{d^2}, d^2 \times \frac{b^2}{d^2}\right) \\
&= d^2 \gcd\left(\frac{a^2}{d^2}, \frac{b^2}{d^2}\right) \\
&= d^2 \times 1 \\
&= d^2.
\end{aligned}
$$

Similarly, we obtain $\gcd\left(x^2, y^2\right) = d^2$.
Therefore, $\gcd\left(a^2, b^2\right) = \gcd\left(x^2, y^2\right)$.

Problem 107. Given a and b are two positive integers. Let x and y be two integers such that $ax + by = m$. If $d = \gcd(a, b)$, prove that $d \mid m$.

Solution. We have $ax + by = m$.
Since $d = \gcd(a, b)$, we obtain $d \mid a$ and $d \mid b$.
It follows that $d \mid ax$ and $d \mid by$.
Then $d \mid ax + by = m$.
Therefore, $d \mid m$.

Problem 108. Given n is a positive integer. Prove that

$$\gcd\left(n! + 1, (n + 1)! + 1\right) = 1.$$

Solution. Let $d = \gcd\left(n! + 1, (n + 1)! + 1\right)$.
Then $d \mid n! + 1$ and $d \mid (n + 1)! + 1$.
It follows that

$$d \mid (n + 1)\left(n! + 1\right) - (n + 1)! - 1$$
$$= (n + 1)! + n + 1 - (n + 1)! - 1$$
$$= n.$$

We obtain

$$d \mid (n + 1)! + 1 - (n + 1)\, n!$$
$$= (n + 1)! + 1 - (n + 1)!$$
$$= 1.$$

It follows that $d = 1$.
Therefore, $\gcd\left(n! + 1, (n + 1)! + 1\right) = 1$.

Problem 109. Given three positive integers a, x, y such that $\gcd(x, y) = 1$ and $a \mid xy$. Prove that $a = bc$ for some positive integers b and c such that $\gcd(b, c) = 1$.

Solution. Let $b = \gcd\left(a, x\right)$. Then $b \mid a$ and $b \mid x$.
Taking $c = \dfrac{a}{b}$ and $d = \dfrac{x}{b}$, then $a = bc$ and $x = bd$. Hence, to prove the given statement, it is sufficient to prove that $c \mid y$.
We have $\gcd\left(c, d\right) = \gcd\left(\dfrac{a}{b}, \dfrac{x}{b}\right) = 1$.
Since $a \mid xy$, then $xy = ak$ for some integers k.
It implies that $bdy = bck$.
We obtain $dy = ck$. It follows that $c \mid dy$.
Then $c \mid y$.
Therefore, $a = bc$, where $\gcd(b, c) = 1$.

Problem 110. Given three positive integers a, b and c. Prove that

1. $\gcd\left(a, b, c\right) = \gcd\left[\gcd\left(a, b\right), c\right]$;

2. $\operatorname{lcm}\left(a, b, c\right) = \operatorname{lcm}\left[\operatorname{lcm}\left(a, b\right), c\right]$.

Solution. Prove that

1. $\gcd(a, b, c) = \gcd[\gcd(a, b), c]$;
   Let $d = \gcd(a, b, c)$ and $d' = \gcd[\gcd(a, b), c]$.
   By definition, we have $d \mid a, d \mid b$ and $d \mid c$. It turns out that $d \mid \gcd(a, b)$. Then d is a common divisor of $\gcd(a, b)$ and c. It follows that

   $$d \mid d'. \tag{1}$$

   Moreover, $d' \mid \gcd(a, b)$ and $d' \mid c$. Then $d' \mid a, d' \mid b$ and $d' \mid c$. It follows that d' is a common divisor of a, b and c. We obtain

   $$d' \mid d. \tag{2}$$

   From (1) and (2), we obtain $d = d'$.
   Therefore, $\gcd(a, b, c) = \gcd[\gcd(a, b), c]$.

2. $\operatorname{lcm}(a, b, c) = \operatorname{lcm}[\operatorname{lcm}(a, b), c]$.
   Let $l = \operatorname{lcm}(a, b, c)$ and $l' = \operatorname{lcm}[\operatorname{lcm}(a, b), c]$.
   By definition, we have $a \mid l, b \mid l$ and $c \mid l$.
   It follows that $\operatorname{lcm}(a, b) \mid l$ and $c \mid l$.
   It turns out that

   $$\operatorname{lcm}[\operatorname{lcm}(a, b), c] \mid l.$$

   We obtain

   $$l' \mid l. \tag{1}$$

   Moreover, $\operatorname{lcm}(a, b) \mid l'$ and $c \mid l'$.
   It follows that $a \mid l', b \mid l'$ and $c \mid l'$.
   We obtain l' is a common multiple of a, b and c.
   It follows that

   $$l \mid l'. \tag{2}$$

   From (1) and (2), we obtain $l = l'$.
   Therefore, $\operatorname{lcm}(a, b, c) = \operatorname{lcm}[\operatorname{lcm}(a, b), c]$.

Problem 111. Given x and y are two positive integers such that $\gcd(x, y) = 1$. Prove that $\operatorname{lcm}(x, y) = xy$.

Problem 112. Given three positive integers a, b and c such that $a \mid c$ and $b \mid c$. Prove that $\operatorname{lcm}(a, b) \mid c$.

Solution. Let $l = \operatorname{lcm}(a, b)$. Suppose that $l \nmid c$.
From division algorithm, we have $c = lq + r$, where $0 < r < l$.
It follows that $r = c - lq$.

By knowing that $a \mid c$ and $b \mid c$, it turns out that $a \mid r$ and $b \mid r$. Hence, r is a common multiple of a and b.
As a result, $l \leq r$, which is a contradiction.
Therefore, $\text{lcm}(a, b) \mid c$.

Problem 113. Given three positive integers a, b and x such that $a \mid x, b \mid x$ and $\gcd(a, b) = 1$. Prove that $ab \mid x$.

Solution. Since $a \mid x$ and $b \mid x$, then $x = am$ and $x = bn$ for some positive integers m and n. It follows that $am = bn$.
Then $a \mid bn$. By knowing that $\gcd(a, b) = 1$, we obtain $a \mid n$. It implies that $n = ak$ for some integers k. Hence, $x = abk$.
Consequently, $ab \mid x$.

Remark 4. The above problem can be proved as the following. From 112, we have $\text{lcm}(a, b) \mid x$. By knowing that $\gcd(a, b) = 1$, it follows that $\text{lcm}(a, b) = ab$. Hence, $ab \mid x$.

Problem 114. Given five positive integers a, b, c, x and y such that $\gcd(a, b) = 1$. Suppose that $a \mid c^x - 1$ and $b \mid c^y - 1$. Prove that

$$ab \mid c^{\text{lcm}(x,y)} - 1.$$

Solution. We know that $\gcd(x, y) \, \text{lcm}(x, y) = xy$.
Then

$$\text{lcm}(x, y) = \frac{xy}{\gcd(x, y)} = mx$$

, where $m = \dfrac{y}{\gcd(x, y)}$.
It follows that

$$
\begin{aligned}
c^{\text{lcm}(x,y)} - 1 &= c^{mx} - 1 \\
&= (c^x)^m - 1 \\
&= (c^x - 1)\left[c^{x(m-1)} + c^{x(m-2)} + \dots + 1\right].
\end{aligned}
$$

Then $c^x - 1 \mid c^{\text{lcm}(x,y)} - 1$. Moreover, $a \mid c^x - 1$.
It turns out that

$$a \mid c^{\text{lcm}(x,y)} - 1. \tag{1}$$

Similarly, we obtain

$$b \mid c^{\text{lcm}(x,y)} - 1. \tag{2}$$

Since $\gcd(a, b) = 1$, from (1) and (2), it implies that

$$ab \mid c^{\text{lcm}(x,y)} - 1.$$

Therefore, $ab \mid c^{\text{lcm}(x,y)} - 1$.

Problem 115. Given three positive integers a, x and y such that $\gcd(x, y) = 1$. Prove that $\gcd(a, xy) = \gcd(a, x) \times \gcd(a, y)$.

Solution. Let $d = \gcd(a, xy)$. Then $d \mid a$ and $d \mid xy$. It turns out that $d = pq$, where $p \mid x, q \mid y$ and $\gcd(p, q) = 1$. We obtain $pq \mid a$. Then $p \mid a$ and $q \mid a$. Hence, p is a common divisor of a and x and q is a common divisor of a and y. As a result, $p \mid \gcd(a, x)$ and $q \mid \gcd(a, y)$. We obtain $pq \mid \gcd(a, x) \times \gcd(a, y)$. That is,

$$d \mid \gcd(a, x) \times \gcd(a, y). \tag{1}$$

Let $d_1 = \gcd(a, x)$ and $d_2 = \gcd(a, y)$. We obtain $d_1 \mid a, d_1 \mid x, d_2 \mid a$ and $d_2 \mid y$. We obtain $d_1 \mid xy$ and $d_2 \mid xy$. Hence, d_1 and d_2 are both common divisors of a and xy. It follows that $d_1 \mid \gcd(a, xy)$ and $d_2 \mid \gcd(a, xy)$. By knowing that $\gcd(x, y) = 1$, then $\gcd(d_1, d_2) = 1$. It follows that $d_1 d_2 \mid \gcd(a, xy)$. Hence,

$$\gcd(a, x) \times \gcd(a, y) \mid d. \tag{2}$$

From (1) and (2), we obtain

$$\gcd(a, xy) = \gcd(a, x) \times \gcd(a, y).$$

Therefore, $\gcd(a, xy) = \gcd(a, x) \times \gcd(a, y)$.

Problem 116. Given a and b are two positive integers. If $\gcd(a, b) = 3$, find the following expressions:

1. $\gcd\left(a^3, b^2\right)$;

2. $\gcd\left(2a^2, b^3\right)$;

3. $\gcd\left(\gcd\left(a^2, b^2\right), 2a^2\right)$;

4. $\gcd\left(\gcd\left(a^2, b^2\right), 3b^3\right)$.

Solution. Find the following expressions:

1. $\gcd\left(a^3, b^2\right)$;
   Since $\gcd(a, b) = 3$, then $a = 3m$ and $b = 3n$, where $\gcd(m, n) = 1$. It follows that

$$\gcd\left(a^3, b^2\right) = \gcd\left(27m^3, 9n^2\right)$$

$$= 9 \gcd \left(3m^3, n^2\right)$$
$$= 9 \gcd \left(3, n^2\right)$$
$$= 9 \times 1 \ \text{ or } \ 9 \times 3$$
$$= 9 \ \text{ or } \ 27.$$

Therefore, $\gcd \left(a^3, b^2\right) = 9$ or 27.

2. $\gcd \left(a^2, b^3\right)$;
   We have

$$\gcd \left(2a^2, b^3\right) = \gcd \left(18m^2, 27n^3\right)$$
$$= 9 \gcd \left(2m^2, 3n^3\right)$$
$$= 9 \gcd \left(2m^2, 3\right)$$
$$= 9 \gcd \left(m^2, 3\right)$$
$$= 9 \times 1 \ \text{ or } \ 9 \times 3$$
$$= 9 \ \text{ or } \ 27.$$

Therefore, $\gcd \left(a^2, b^3\right) = 9$ or 27.

3. $\gcd \left(\gcd \left(a^2, b^2\right), 2a^2\right)$;
   We have $\gcd \left(a^2, b^2\right) = \gcd \left(9m^2, 9n^2\right) = 9 \gcd \left(m^2, n^2\right) = 9$.
   It follows that

$$\gcd \left(\gcd \left(a^2, b^2\right), 2a^2\right) = \gcd \left(9, 2a^2\right)$$
$$= \gcd \left(9, 2 \times 9m^2\right)$$
$$= 9 \gcd \left(1, 2m^2\right)$$
$$= 9 \times 1$$
$$= 9.$$

Therefore, $\gcd \left(\gcd \left(a^2, b^2\right), 2a^2\right) = 9$.

4. $\gcd \left(\gcd \left(a^2, b^2\right), 3b^3\right)$.
   From the above computation, we have $\gcd \left(a^2, b^2\right) = 9$.
   We obtain

$$\gcd \left(\gcd \left(a^2, b^2\right), 3b^2\right) = \gcd \left(9, 3 \times 9n^2\right)$$
$$= \gcd \left(9, 27n^2\right)$$
$$= 9 \gcd \left(1, 3n^2\right)$$

$$= 9 \times 1$$
$$= 9.$$

Therefore, $\gcd\left(\gcd\left(a^2, b^2\right), 3b^3\right) = 9$.

Problem 117. Given two positive integers a and b. Let d be a positive odd number such that $d \mid a + b$ and $d \mid a - b$. Prove that

$$d \mid \gcd(a, b).$$

Solution. Since $d \mid a + b$ and $d \mid a - b$, it follows that

$$d \mid (a + b) + (a - b) = 2a$$

and

$$d \mid (a + b) - (a - b) = 2b.$$

Hence, d is a common divisor of $2a$ and $2b$.
We obtain $d \mid \gcd(2a, 2b) = 2\gcd(a, b)$.
Moreover, d is an odd number. Then $\gcd(d, 2) = 1$.
It turns out that $d \mid \gcd(a, b)$.
Therefore, $d \mid \gcd(a, b)$.

Problem 118. Given positive integers a, b, m, n, x and y, where b is a prime number. Suppose that $d \mid ax + by, d \mid am + bn$ and $my - nx = b$. Find the possible values of $\gcd(ax + by, am + bn)$.

Solution. We have

$$d \mid m(ax + by) - x(am + bn)$$
$$= amx + bmy - amx - bnx$$
$$= b(my - nx).$$

Since $my - nx = b$, it follows that $d \mid b^2$.
By knowing that b is a prime number, it follows that $d = 1, b$ or b^2.
Therefore, $d = 1, b$ or b^2.

Problem 119. Given n and d are two positive integers such that $d \mid 3n + 1$ and $d \mid 5n + 2$. Find d.

Solution. We have $d \mid 3n + 1$ and $d \mid 5n + 2$.
Then

$$d \mid 5(3n + 1) - 3(5n + 2)$$

$$= 15n + 5 - 15n - 6$$
$$= -1.$$

It follows that $d = 1$.
Therefore, $d = 1$.

Problem 120. Given n and d are two positive integers such that $d \mid 5n - 2$ and $d \mid 4n - 3$. Find d.

Solution. We have $d \mid 5n - 2$ and $d \mid 4n - 3$.
Then

$$d \mid 4(5n - 2) - 5(4n - 3)$$
$$= 20n - 8 - 20n + 15$$
$$= 7.$$

It follows that $d = 1$ or $d = 7$.
Therefore, $d = 1$ or $d = 7$.

Problem 121. Given a and b are positive integers such that $\gcd(a, b) = 1$ and ab is a perfect square. Prove that a and b are perfect squares.

Solution. Since ab is a perfect square, then

$$ab = (p_1^{\alpha_1} p_2^{\alpha_2} ... p_n^{\alpha_n})^2 = p_1^{2\alpha_1} p_2^{2\alpha_2} ... p_n^{2\alpha_n}$$

, where $p_1, p_2, ..., p_n$ are prime numbers and $\alpha_1, \alpha_2, ..., \alpha_n$ are non-negative integers. By knowing that $\gcd(a, b) = 1$, it follows that

$$a = q_1^{2\beta_1} q_2^{2\beta_2} ... q_i^{2\beta_i}$$

and

$$b = q_{i+1}^{2\beta_{i+1}} q_{i+2}^{2\beta_{i+2}} ... q_n^{2\beta_n}$$

, where $\{q_1, q_2, ..., q_n\}$ is a permutation of $\{p_1, p_2, ..., p_n\}$.
It implies that

$$a = \left(q_1^{\beta_1} q_2^{\beta_2} ... q_i^{\beta_i} \right)^2$$

and

$$b = \left(q_{i+1}^{\beta_{i+1}} q_{i+2}^{\beta_{i+2}} ... q_n^{\beta_n} \right)^2.$$

Therefore, a and b are perfect squares.

Problem 122. Given x, y and z are three integers such that $\gcd(x, y) = \gcd(y, z) = \gcd(z, x) = 1$. Prove that

$$\gcd(x, y, z) \times \operatorname{lcm}(x, y, z) = xyz.$$

Solution. Since $\gcd(x, y) = \gcd(y, z) = \gcd(z, x) = 1$, we obtain

$$\begin{aligned}
\gcd(x, y, z) &= \gcd(\gcd(x, y), z) \\
&= \gcd(1, z) \\
&= 1.
\end{aligned}$$

Moreover,

$$\begin{aligned}
\operatorname{lcm}(x, y, z) &= \operatorname{lcm}(\operatorname{lcm}(x, y), z) \\
&= \operatorname{lcm}(xy, z) \\
&= xyz.
\end{aligned}$$

It follows that $\gcd(x, y, z) \times \operatorname{lcm}(x, y, z) = xyz$.

Problem 123. Given two positive integers a and b such that $\gcd(a, b) = 1$. Find the possible values of the following expressions:

1. $\gcd(a, a + 1, a + 2)$;

2. $\gcd(a, (a + 1)(a + 2))$;

3. $\gcd(a, (a + 2)(a, +3))$;

4. $\gcd(a(a + 1), (a + 2)(a + 3))$;

5. $\gcd(a^2 - b^2, a^2 + b^2)$;

6. $\gcd(2a - 2b, a^3 - b^3)$;

7. $\gcd(a^2 - b^2, a^3 - b^3)$.

Solution. Find the possible values of the following expressions:

1. $\gcd(a, a + 1, a + 2)$;
   We have already known that two consecutive integers are relatively prime. It follows that

   $$\begin{aligned}
   \gcd(a, a + 1, a + 2) &= \gcd(\gcd(a, a + 1), a + 2) \\
   &= \gcd(1, a + 2) \\
   &= 1.
   \end{aligned}$$

   Therefore, $\gcd(a, a + 1, a + 2) = 1$.

2. $\gcd(a, (a+1)(a+2))$;

   We know that given three positive integers a, x and y such that $\gcd(x, y) = 1$, then $\gcd(a, xy) = \gcd(a, x) \times \gcd(a, y)$. Since $a+1$ and $a+2$ are two consecutive positive integers, they are relatively prime. It follows that

   $$\gcd(a, (a+1)(a+2)) = \gcd(a, a+1) \times \gcd(a, a+2).$$

   We have $\gcd(a, a+1) = 1$.
   Let $d = \gcd(a, a+2)$. Then $d \mid a$ and $d \mid a+2$. It follows that $d \mid a+2 - a = 2$. Hence, $d = 1$ or $d = 2$. That is,

   $$\gcd(a, a+2) = 1$$

   or

   $$\gcd(a, a+2) = 2.$$

   Therefore, $\gcd(a, (a+1)(a+2)) = 1$ or 2.

3. $\gcd(a, (a+2)(a, +3))$;

   Since $a+2$ and $a+3$ are relatively prime, it follows that

   $$\gcd(a, (a+2)(a+3)) = \gcd(a, a+2) \times \gcd(a, a+3).$$

   From the above proof, we have $\gcd(a, +2) = 1$ or 2.
   Similarly, $\gcd(a, a+3) = 1$ or 3.
   We obtain $\gcd(a, (a+2)(a+3)) = 1, 2, 3$ or 6.

4. $\gcd(a(a+1), (a+2)(a+3))$;

   Since a and $a+1$ are relatively prime and $a+2$ and $a+3$ are relatively prime, we obtain

   $$\gcd(a(a+1), (a+2)(a+3))$$
   $$= \gcd(a, (a+2)(a+3)) \times \gcd(a+1, (a+2)(a+3))$$
   $$= \gcd(a, a+2) \times \gcd(a, a+3) \times \gcd(a+1, a+2)$$
   $$\times \gcd(a+1, a+3).$$

   We know that

   - $a+1$ and $a+2$ are two consecutive positive integers. Then $\gcd(a+1, a+2) = 1$.
   - $\gcd(a, a+2) = 1$ or 2.
   - $\gcd(a, a+3) = 1$ or 3.

- $\gcd(a+1, a+3) = 1$ or 2.

Therefore, $\gcd\left(a\left(a+1\right), \left(a+2\right)\left(a+3\right)\right) = 1, 2, 3, 4, 6$ or 12.

5. $\gcd\left(a^2 - b^2, a^2 + b^2\right)$;
   Let $d = \gcd\left(a^2 - b^2, a^2 + b^2\right)$.
   We obtain $d \mid a^2 - b^2$ and $d \mid a^2 + b^2$.
   It follows that $d \mid \left(a^2 - b^2\right) + \left(a^2 + b^2\right) = 2a^2$.
   Similarly, we obtain $d \mid 2b^2$.
   It implies that d is a common divisor of $2a^2$ and $2b^2$.
   Hence,
   $$d \mid \gcd\left(2a^2, 2b^2\right) = 2[\gcd\left(a, b\right)]^2 = 2.$$
   Therefore, $\gcd\left(a^2 - b^2, a^2 + b^2\right) = 1$ or 2.

6. $\gcd\left(2a - 2b, a^3 - b^3\right)$;
   We have
   $$\gcd\left(2a - 2b, a^3 - b^3\right) = \gcd\left(2\left(a - b\right), \left(a - b\right)\left(a^2 + ab + b^2\right)\right)$$
   $$= \left(a - b\right)\gcd\left(2, a^2 + ab + b^2\right).$$

   Let $d = \gcd\left(2, a^2 + ab + b^2\right)$.
   It follows that $d \mid 2$. Then $d = 1$ or 2.
   Therefore, $\gcd\left(2a - 2b, a^3 - b^3\right) = a - b$ or $2(a - b)$.

7. $\gcd\left(a^2 - b^2, a^3 - b^3\right)$.
   We have
   $$\gcd\left(a^2 - b^2, a^3 - b^3\right)$$
   $$= \gcd\left(\left(a - b\right)\left(a + b\right), \left(a - b\right)\left(a^2 + ab + b^2\right)\right)$$
   $$= \left(a - b\right)\gcd\left(a + b, a^2 + ab + b^2\right).$$

   Let $d = \gcd(a + b, a^2 + ab + b^2)$.
   Then $d \mid a + b$ and $d \mid a^2 + ab + b^2$.
   We obtain
   $$d \mid \left(a^2 + ab + b^2\right) - a\left(a + b\right) = b^2.$$

   Similarly, $d \mid a^2$.
   It follows that d is a common divisor of a^2 and b^2.
   Then $d \mid \gcd(a^2, b^2) = [\gcd(a, b)]^2 = 1$.
   We obtain $d = 1$.
   Therefore, $\gcd\left(a^2 - b^2, a^3 - b^3\right) = a - b$.

Problem 124. For all positive integers a, prove that

$$\gcd\left(a^{2^n}+1, a^{2^m}+1\right) = 1 \ \text{ or } 2$$

for all positive integers m and n.

Solution. Without loss of generality, suppose that $n < m$. It follows that $a^{2^n}+1$ divides $a^{2^m}-1$. That is, there exists an integer k such that $a^{2^m}-1 = k\left(a^{2^n}+1\right)$. Let $d = \gcd\left(a^{2^n}+1, a^{2^m}+1\right)$. Then $d \mid a^{2^n}+1$ and $d \mid a^{2^m}+1$. Using the property of divisibility, we obtain

$$d \mid a^{2m}+1 - k\left(a^{2^n}+1\right)$$
$$= a^{2m}-1+2 - k\left(a^{2^n}+1\right)$$
$$= k\left(a^{2^n}+1\right)+2 - k\left(a^{2^n}+1\right)$$
$$= 2.$$

Hence, $d = 1$ or 2.
Therefore, $\gcd\left(a^{2^n}+1, a^{2^m}+1\right) = 1$ or 2.

Problem 125. Given a positive integer n. Prove that

$$\frac{\gcd\left(n\left(n+1\right)\left(n+2\right), 3n+9\right)}{\gcd\left(n\left(n+2\right),\left(n+1\right)\left(n+3\right)\right)}$$

is an integer.

Solution. Since $n, n+1$ and $n+2$ are three consecutive positive integers, then their product is divisible by 3. That is,

$$3 \mid n(n+1)(n+2).$$

Moreover, $3n+9 = 3(n+3)$. It follows that $3 \mid 3n+9$. It turns out that 3 is a common multiple of $n(n+1)(n+2)$ and $3n+9$. We obtain $3 \mid \gcd\left(n\left(n+1\right)\left(n+2\right), 3n+9\right)$.
We know that two consecutive integers are relatively prime. Hence, $\gcd(n, n+1) = 1, \gcd(n+1, n+2) = 1$ and $\gcd(n+2, n+3) = 1$. We obtain

$$\gcd\left(n\left(n+2\right),\left(n+1\right)\left(n+3\right)\right) = \gcd\left(n,\left(n+1\right)\left(n+3\right)\right)$$

$$= \gcd(n, n+3).$$

Let $d = \gcd(n, n+3)$. Then $d \mid n$ and $d \mid n+3$.
It follows that $d \mid (n+3) - n = 3$.
We obtain $d = 1$ or $d = 3$.
Hence, $\gcd(n(n+2), (n+1)(n+3)) = 1$ or 3.
That is, $\gcd(n(n+2), (n+1)(n+3)) \mid 3$.
It turns out that

$$\gcd(n(n+2), (n+1)(n+3)) \mid \gcd(n(n+1)(n+2), 3n+9).$$

Therefore, $\dfrac{\gcd(n(n+1)(n+2), 3n+9)}{\gcd(n(n+2), (n+1)(n+3))}$ is an integer.

Problem 126. Given a positive integer n. Simplify

$$F = \frac{\gcd(n(n+1), (n+1)(n+3))}{\gcd\left(n(n+1), (n+1)^2\right)}.$$

Solution. We have

$$\gcd(n(n+1), (n+1)(n+3)) = (n+1)\gcd(n, n+3)$$

and

$$\gcd\left(n(n+1), (n+1)^2\right) = (n+1)\gcd(n, n+1) = n+1.$$

It follows that

$$F = \frac{(n+1)\gcd(n, n+3)}{n+1}$$
$$= \gcd(n, n+3).$$

Hence, $F \mid n$ and $F \mid n+3$.
Then $F \mid n+3 - n = 3$. It implies that $F = 1$ or 3.
Therefore, $F = 1$ or 3.

Problem 127. Let n be an even number. Find all positive integers a and b such that $\gcd(a, b) = 1$ and $a + b \mid a^n + b^n$.

Solution. Since n is an even number, then

$$a^n - b^n = \left(a^2\right)^{\frac{n}{2}} - \left(b^2\right)^{\frac{n}{2}}$$

93

$$= \left(a^2 - b^2\right)\left(a^{2n-2} + a^{2n-4} + \dots + b^{2n-2}\right)$$
$$= (a+b)(a-b)\left(a^{2n-2} + a^{2n-4} + \dots + b^{2n-2}\right).$$

Then $a + b \mid a^n - b^n$.

Moreover, $a + b \mid a^n + b^n$.

Using the property of divisibility, we obtain

$$a + b \mid (a^n - b^n) + (a^n + b^n) = 2a^n.$$

Similarly, we obtain

$$a + b \mid (a^n + b^n) - (a^n - b^n) = 2b^n.$$

Hence, $a + b$ is a common divisor of $2a^n$ and $2b^n$.

It turns out that $a + b \mid \gcd(2a^2, 2b^n)$.

Since $\gcd(a, b) = 1$, then

$$\gcd\left(2a^n, 2b^n\right) = 2\gcd\left(a^n, b^n\right)$$
$$= 2[\gcd\left(a, b\right)]^n$$
$$= 2.$$

We obtain $a + b \mid 2$. Hence, $a = b = 1$.

Therefore, $a = b = 1$.

Problem 128. Given a and b are two distinct positive integers. Prove that

$$|a - b| \geq \gcd(a, b).$$

Solution. Let $d = \gcd(a, b)$. Then there exist integers x and y such that $a = dx$ and $b = dy$, where $\gcd(x, y) = 1$.

It follows that

$$(a - b)^2 = (dx - dy)$$
$$= d^2(x - y)^2.$$

Since $a \neq b$, then $x \neq y$. We obtain $(x - y)^2 \geq 1$. The equality holds if and only if $x - y = 1$. Hence, $(a - b)^2 \geq d^2$. Thus, $|a - b| \geq d$. Therefore, $|a - b| \geq \gcd(a, b)$.

Solution. Let $l = \text{lcm}(x, y)$. Since xy is a common multiple of x and y, by definition, we obtain

$$l \leq xy. \tag{1}$$

Moreover, $x \mid l$ and $y \mid l$. By knowing that $\gcd(x, y) = 1$, we obtain

$$xy \mid l.$$

It turns out that

$$xy \leq l. \tag{2}$$

From (1) and (2), we obtain $\operatorname{lcm}(x, y) = xy$.

Problem 129. Given that a and b are two positive integers. Prove that

$$\operatorname{lcm}(a, b) \times \gcd(a, b) = ab.$$

Solution. Let $d = \gcd(a, b)$ and $l = \operatorname{lcm}(a, b)$. Then $a = dx$ and $b = dy$, where $\gcd(x, y) = 1$. It follows that

$$
\begin{aligned}
ab &= d^2 xy \\
&= \gcd(a, b) \times (dxy) \\
&= \gcd(a, b) \times d \operatorname{lcm}(x, y) \\
&= \gcd(a, b) \times \operatorname{lcm}(dx, dy) \\
&= \gcd(a, b) \times \operatorname{lcm}(a, b).
\end{aligned}
$$

Therefore, $\operatorname{lcm}(a, b) \times \gcd(a, b) = ab$.

Problem 130. Given a and b are two positive integers such that $a > b$. Prove that

$$\frac{1}{\operatorname{lcm}(a, b)} \leq \frac{1}{b} - \frac{1}{a}.$$

Solution. For all positive integers a and b, we have

$$|a - b| \geq \gcd(a, b).$$

Since $a > b$ and $\gcd(a, b) = \dfrac{ab}{\operatorname{lcm}(a, b)}$, we obtain

$$a - b \geq \frac{ab}{\operatorname{lcm}(a, b)}.$$

Then

$$\frac{1}{\operatorname{lcm}(a, b)} \leq \frac{a - b}{ab} = \frac{1}{b} - \frac{1}{a}.$$

Therefore, $\dfrac{1}{\operatorname{lcm}(a, b)} \leq \dfrac{1}{b} - \dfrac{1}{a}$.

Problem 131. Given $x_1, x_2, ..., x_n$ is a strictly increasing sequence of positive integers. Prove that

$$\sum_{k=1}^{n-1} \frac{1}{\text{lcm}\,(x_k, x_{k+1})} \leq 1 - \frac{1}{x_n}.$$

Solution. Since $x_1, x_2, ..., x_n$ is a strictly increasing sequence of positive integers, we obtain $1 \leq x_1 < x_2 < ... < x_n$.
We know that

$$\frac{1}{\text{lcm}\,(x_k, x_{k+1})} \leq \frac{1}{x_k} - \frac{1}{x_{k+1}}.$$

It follows that

$$\sum_{k=1}^{n-1} \frac{1}{\text{lcm}\,(x_k, x_{k+1})} \leq \sum_{k=1}^{n-1} \left(\frac{1}{x_k} - \frac{1}{x_{k+1}} \right)$$

$$= \frac{1}{x_1} - \frac{1}{x_n}$$

$$\leq 1 - \frac{1}{x_n}.$$

Therefore, $\displaystyle\sum_{k=1}^{n-1} \frac{1}{\text{lcm}\,(x_k, x_{k+1})} \leq 1 - \frac{1}{x_n}$.

Problem 132. Given two positive integers a and b such that $a > b$ and $\dfrac{\text{lcm}\,(a, b)}{\gcd\,(a, b)} = a - b$. Prove that ab is a perfect cube.

Solution. Let $d = \gcd(a, b)$. Then $a = dx$ and $b = dy$, where $\gcd(x, y) = 1$. Hence, $\text{lcm}\,(a, b) = dxy$. From $\dfrac{\text{lcm}\,(a, b)}{\gcd\,(a, b)} = a - b$, we obtain

$$\frac{dxy}{d} = dx - dy.$$

Then

$$xy = d\,(x - y). \tag{1}$$

From $xy = dx - dy$, we obtain $x \mid dy$. Then $x \mid d$ since $\gcd(x, y) = 1$. Similarly, we obtain $y \mid d$. Consequently, $xy \mid d$. From (1), we obtain $xy = d$ and $x - y = 1$. It turns out that

$$ab = (dx)\,(dy)$$

$$= d^2xy$$
$$= d^3$$
$$= [\gcd(a,b)]^3.$$

Therefore, ab is a perfect cube.

Problem 133. For all positive integers n, prove that

$$\frac{n!}{(x+1)(x+2)\dots(x+n)} = \sum_{k=1}^{n} \frac{(-1)^{k-1}kC(n,k)}{x+k}.$$

Solution. Suppose that

$$\frac{n!}{(x+1)(x+2)\dots(x+n)} = \frac{a_1}{x+1} + \frac{a_2}{x+2} + \dots + \frac{a_n}{x+n}.$$

Multiply both sides of the equality by $(x+1)(x+2)\dots(x+n)$ and taking $x = -i$, we obtain

$$n! = (-i+1)(-i+2) \times \dots \times 1 \times 2 \times \dots \times (-i+n)\,a_i$$
$$= (-1)^{i-1}(i-1)(i-2) \times \dots \times 1 \times 2 \times \dots \times (n-i)\,a_i$$
$$= (-1)^{i-1}(i-1)!\,(n-1)!a_i$$
$$= (-1)^{i-1} \times \frac{i!\,(n-i)!}{i}a_i$$

It follows that

$$a_i = (-1)^{i-1}\frac{i \times n!}{i!\,(n-i)!}$$
$$= (-1)^{i-1}iC(n,i).$$

Therefore, $\dfrac{n!}{(x+1)(x+2)\dots(x+n)} = \displaystyle\sum_{k=1}^{n} \frac{(-1)^{k-1}kC(n,k)}{x+k}.$

Problem 134. Given two positive integers x and y. Prove that $yC(x+y,y)$ divides $\operatorname{lcm}(x+1, x+2, \dots, x+y)$.

Solution. We have

$$C(x+y,x) = \frac{(x+y)!}{x!y!}$$

$$= \frac{(x+1)(x+2)\dots(x+y)}{y!}.$$

Then $\dfrac{1}{C(x+y,x)} = \dfrac{y!}{(x+1)(x+2)\dots(x+y)}$.

From the previous problem, we have

$$\frac{1}{C(x+y,x)} = \sum_{k=1}^{y} \frac{(-1)^{k-1} k C(y,k)}{x+k}$$

$$= \sum_{k=1}^{y} \frac{(-1)^{k-1} y C(y-1,k-1)}{x+k}$$

$$= y \sum_{k=1}^{y} \frac{(-1)^{k-1} C(y-1,k-1)}{x+k}.$$

Then

$$\frac{1}{yC(x+y,x)} = \sum_{k=1}^{y} \frac{(-1)^{k-1} C(y-1,k-1)}{x+k}.$$

We see that the right-hand side of the last equality is of the form

$$\frac{m}{\mathrm{lcm}(x+1, x+2, \dots, x+y)}$$

, where m is an integer. It turns out that

$$\frac{1}{yC(x+y,x)} = \frac{m}{\mathrm{lcm}(x+1, x+2, \dots, x+y)}.$$

Then

$$m = \frac{\mathrm{lcm}(x+1, x+2, \dots, x+y)}{yC(x+y,x)}.$$

Therefore, $yC(x+y,y)$ divides $\mathrm{lcm}(x+1, x+2, \dots, x+y)$.

Problem 135. Given $x_1, x_2, \dots, x_n$ are n positive integers. Prove that

$$\mathrm{lcm}(x_1, x_2, \dots, x_n) \geq \frac{x_1 + x_2 + \dots + x_n}{n}.$$

Solution. By definition, we obtain

$$\mathrm{lcm}(x_1, x_2, \dots, x_n) \geq x_1$$
$$\mathrm{lcm}(x_1, x_2, \dots, x_n) \geq x_2$$

$$\vdots$$

and $\quad \operatorname{lcm}(x_1, x_2, ..., x_n) \geq x_n.$

Adding all of the above inequalities, we obtain

$$n \operatorname{lcm}(x_1, x_2, ..., x_n) \geq x_1 + x_2 + ... + x_n.$$

Therefore, $\operatorname{lcm}(x_1, x_2, ..., x_n) \geq \dfrac{x_1 + x_2 + ... + x_n}{n}.$

Problem 136. For all positive integers n greater than 1, prove that

$$\frac{\operatorname{lcm}(1, 2, ..., n+1)}{n+1} = \operatorname{lcm}(C(n,0), C(n,1), ..., C(n,n)).$$

Solution.

$$\frac{\operatorname{lcm}(1, 2, ..., n+1)}{n+1} = \operatorname{lcm}(C(n,0), C(n,1), ..., C(n,n))$$

is equivalent to

$$\operatorname{lcm}(1, 2, ..., n+1) = (n+1)\operatorname{lcm}(C(n,0), C(n,1), ..., C(n,n)).$$

We have

$$
\begin{aligned}
(n+1)C(n,i) &= (n+1) \times \frac{n!}{i!(n-i)!} \\
&= (i+1) \times \frac{(n+1)!}{(i+1)! \times (n-i)!} \\
&= (i+1)C(n+1, i+1).
\end{aligned}
$$

From Problem 134, we have $(n+1)C(n,i)$ divides

$$\operatorname{lcm}(n+1-i, n+2-i, ..., n+1).$$

Hence,

$$(n+1)C(n,i) \mid \operatorname{lcm}(1, 2, ..., n+1).$$

Then

$$(n+1)\operatorname{lcm}(C(n,0), C(n,1), ..., C(n,n)) \mid \operatorname{lcm}(1, 2, ..., n+1). \tag{1}$$

99

For all integers i such that $0 \leq i \leq n$, we have

$$(n+1) \, C \, (n, i) = (i+1) \, C \, (n+1, i+1) \, .$$

It turns out that $i+1$ divides $(n+1) \, C \, (n, i)$ for all $0 \leq i \leq n$. It follows that $i+1$ divides

$$(n+1) \, \mathrm{lcm} \, (C \, (n, 0) \, , C \, (n, 1) \, , ..., C \, (n, n))$$

for all $0 \leq i \leq n$.
Consequently,

$$\mathrm{lcm}(1, 2, ..., n+1) \mid (n+1) \, \mathrm{lcm} \, (C \, (n, 0) \, , C \, (n, 1) \, , ..., C \, (n, n)) \, . \tag{2}$$

From (1) and (2), we obtain

$$\mathrm{lcm} \, (1, 2, ..., n+1) = (n+1) \, \mathrm{lcm} \, (C \, (n, 0) \, , C \, (n, 1) \, , ..., C \, (n, n)) \, .$$

Therefore,

$$\frac{\mathrm{lcm} \, (1, 2, ..., n+1)}{n+1} = \mathrm{lcm} \, (C \, (n, 0) \, , C \, (n, 1) \, , ..., C \, (n, n)) \, .$$

Problem 137. For all positive integers $n \geq 2$, prove that

$$\mathrm{lcm} \, (1, 2, 3, ..., n) \geq 2^{n-1} \, .$$

Solution. From Problem 135 and Problem 136 , we have

$$\mathrm{lcm} \, (1, 2, ..., n+1) = (n+1) \, \mathrm{lcm} \, (C \, (n, 0) \, , C \, (n, 1) \, , ..., C \, (n, n))$$

$$\geq (n+1) \times \frac{\sum\limits_{k=0}^{n} C \, (n, k)}{n+1}$$

$$= (n+1) \times \frac{2^n}{n+1}$$

$$= 2^n$$

for all $n \geq 2$.
Hence, $\mathrm{lcm} \, (1, 2, ..., n) \geq 2^{n-1}$ for all $n \geq 3$.
For the case $n = 2$, we have

$$\mathrm{lcm} \, (1, 2) \geq 2^0$$

or

$$2 \geq 1, \text{true.}$$

Therefore, $\mathrm{lcm} \, (1, 2, 3, ..., n) \geq 2^{n-1}$ for all $n \geq 2$.

Problem 138. Given three positive integers x, y and z. Prove that $\gcd(x, yz)$ divides $\gcd(x, y) \times \gcd(x, z)$.

Solution. Let $d = \gcd(x, y)$. Then $x = da$ and $y = db$ for some positive integers a and b such that $\gcd(a, b) = 1$. Hence,

$$\gcd(x, yz) \mid \gcd(x, y) \times \gcd(x, z)$$

is equivalent to

$$\gcd(a, bz) \mid \gcd(x, z).$$

Since $\gcd(a, bz) \mid a$ and $\gcd(a, b) = 1$, it turns out that $\gcd(a, bz)$ and b are relatively prime. We know that $\gcd(a, bz) \mid bz$. It follows that $\gcd(a, bz) \mid z$. Moreover, $\gcd(a, bz) \mid a$ and $a \mid x$. Then $\gcd(a, bz) \mid x$. Consequently, $\gcd(a, bz) \mid \gcd(x, z)$. Therefore, $\gcd(x, yz)$ divides $\gcd(x, y) \times \gcd(x, z)$.

Remark 5. From the previous problem, we have $\gcd(x, yz)$ divides $\gcd(x, y) \times \gcd(x, z)$. Since $\gcd(x, y), \gcd(x, z)$ and $\gcd(x, yz)$ are positive integers, we obtain

$$\gcd(x, yz) \le \gcd(x, y) \times \gcd(x, z).$$

This inequality will be used to prove the next problem.

Problem 139. Given $x_1, x_2, \ldots, x_n$ are n positive integers. Prove that

$$\gcd(x_1 x_2 \ldots x_{n-1}, x_n) \le \prod_{k=1}^{n-1} \gcd(x_k, x_n)$$

for all $n \ge 3$.

Solution. We will prove the given statement by induction. For $n = 3$, we have

$$\gcd(x_1 x_2, x_3) \le \gcd(x_1, x_3) \times \gcd(x_2, x_3)$$

, which is true from the above problem.

Suppose that $\gcd(x_1 x_2 \ldots x_{n-1}, x_n) \le \prod_{k=1}^{n-1} \gcd(x_k, x_n)$.

We shall show that

$$\gcd(x_1 x_2 \ldots x_n, x_{n+1}) \le \prod_{k=1}^{n} \gcd(x_k, x_{n+1}).$$

We have

$$\gcd(x_1 x_2 ... x_n, x_{n+1}) = \gcd((x_1 x_2 ... x_{n-1}) x_n, x_{n+1})$$

$$\leq \gcd(x_1 x_2 ... x_{n-1}, x_{n+1}) \times \gcd(x_n, x_{n+1})$$

$$\leq \prod_{k=1}^{n-1} \gcd(x_k x_{n+1}) \times \gcd(x_n, x_{n+1})$$

$$= \prod_{k=1}^{n} \gcd(x_k x_{n+1}).$$

Therefore, $\gcd(x_1 x_2 ... x_{n-1}, x_n) \leq \prod_{k=1}^{n-1} \gcd(x_k, x_n)$ for all $n \geq 3$.

Problem 140. Given n positive integers $x_1, x_2, ..., x_n$. Prove that

$$\operatorname{lcm}(x_1, x_2, ..., x_n) \geq \frac{x_1 x_2 ... x_n}{\prod_{1 \leq i < j \leq n} \gcd(x_i, x_j)}.$$

Solution. We will prove the given inequality by induction. For $n = 2$, we have

$$\operatorname{lcm}(x_1, x_2) \geq \frac{x_1 x_2}{\gcd(x_1, x_2)}$$

, which is true from the fact that

$$\operatorname{lcm}(x_1, x_2) \times \gcd(x_1, x_2) = x_1 x_2.$$

Suppose that $\operatorname{lcm}(x_1, x_2, ..., x_n) \geq \dfrac{x_1 x_2 ... x_n}{\prod_{1 \leq i < j \leq n} \gcd(x_i, x_j)}.$

We shall show that $\operatorname{lcm}(x_1, x_2, ..., x_{n+1}) \geq \dfrac{x_1 x_2 ... x_{n+1}}{\prod_{1 \leq i < j \leq n+1} \gcd(x_i, x_j)}.$

From the inductive hypothesis, we have

$$\operatorname{lcm}(x_1, x_2, ..., x_{n+1})$$

$$= \frac{x_{n+1} \operatorname{lcm}(x_1, x_2, ..., x_n)}{\gcd(\operatorname{lcm}(x_1, x_2, ..., x_{n+1}), x_{n+1})}$$

$$\geq \frac{x_1 x_2 ... x_{n+1}}{\gcd(\operatorname{lcm}(x_1, x_2, ..., x_{n+1}), x_{n+1}) \prod_{1 \leq i < j \leq n} \gcd(x_i, x_n)}.$$

Hence, to prove the statement, it is sufficient to prove that

$$\frac{x_1 x_2 ... x_{n+1}}{\gcd\left(\operatorname{lcm}\left(x_1, x_2, ..., x_{n+1}\right), x_{n+1}\right) \prod\limits_{1 \leq i < j \leq n} \gcd\left(x_i, x_n\right)}$$

$$\geq \frac{x_1 x_2 ... x_{n+1}}{\prod\limits_{1 \leq i < j \leq n+1} \gcd\left(x_i, x_j\right)}.$$

The last inequality is equivalent to

$$\gcd\left(\operatorname{lcm}\left(x_1, x_2, ..., x_n\right), x_{n+1}\right) \leq \prod_{k=1}^{n} \gcd\left(x_k, x_{n+1}\right). \qquad (1)$$

Since $\operatorname{lcm}(x_1, x_2, ..., x_n) \mid x_1 x_2 ... x_n$, then

$$\gcd\left(\operatorname{lcm}\left(x_1, x_2, ..., x_n\right), x_{n+1}\right) \leq \gcd\left(x_1 x_2 ... x_n, x_{n+1}\right)$$

$$\leq \prod_{k=1}^{n} \gcd\left(x_k, x_{n+1}\right).$$

Consequently, (1) is true.

Therefore, $\operatorname{lcm}\left(x_1, x_2, ..., x_n\right) \geq \dfrac{x_1 x_2 ... x_n}{\prod\limits_{1 \leq i < j \leq n} \gcd\left(x_i, x_j\right)}.$